SpringerBriefs in Well-Being and Quality of Life Research

For further volumes:
http://www.springer.com/series/10150

Ronald E. Anderson

Human Suffering and Quality of Life

Conceptualizing Stories and Statistics

Springer

Ronald E. Anderson
University of Minnesota
Wayzata, MN
USA

ISSN 2211-7644 ISSN 2211-7652 (electronic)
ISBN 978-94-007-7668-5 ISBN 978-94-007-7669-2 (eBook)
DOI 10.1007/978-94-007-7669-2
Springer Dordrecht Heidelberg New York London

Library of Congress Control Number: 2013949247

Printed on acid-free paper

Springer is part of Springer Science+Business Media (www.springer.com)

Preface

The original intent of this little book is to take on four very big challenges: (1) a framework that makes it easier to think about suffering and measure it, (2) a compilation of available data on how much suffering exists in the world, (3) rationales for why people should become more aware of the vast volume of severe suffering around the world, and (4) justification for giving higher priority to the reduction of suffering in our personal, state, and global policy objectives. With these goals, you should not be surprised that the book looks at suffering from many different angles.

After 40 years of teaching sociology and research strategies at the University of Minnesota, I retired in order to just do research, travel, and volunteer work. Making this major life change forced me to confront questions of meaning, especially: What makes life worth living? What is the meaning of suffering? What can one do in later life to optimally contribute to ultimate concerns of human beings? My first major conclusion was that compassion is most needed to secure the human race. But after focusing on compassion for several years, I came to realize that the efficacy of compassion is constrained by the huge supply of suffering in the world, which only seems to be expanding.

When I started reading what others had learned about suffering, I discovered a void of knowledge and concluded that doing pioneering research on suffering would be the best way I could use my talents and experience. It has been exciting to discover some elements of suffering, which are as old as human consciousness itself.

This brief book of about 125 pages, follows the structure and format of all SpringerBriefs, of which there are thousands. In the SpringerBrief model, each chapter is like a separate article with its own abstract, keywords, footnotes, and references. This requirement, I believe, is a good one because it forces the author to make each chapter convey a complete statement of its own, but at the same time makes the chapters flow together so that the entire set is an integral whole.

The first chapter begins by explicitly defining some very different types of suffering from which a taxonomy emerged. People think about suffering in very different ways, depending upon their backgrounds in religion, local culture, and unique personal experiences. Chapter 1 discusses eight 'frames for suffering' and Chap. 2 supplies stories for each way of thinking about suffering. How suffering shapes peoples' quality of life becomes clearer through these stories.

Statistics offer only fleeting glimpses of the distress and agony suffered by some in the course of everyday life. But in Chap. 3, you will see how our taxonomy of suffering helps organize and add meaning to statistics on the health of American adults. People react differently to suffering, depending upon whether it is primarily pain, depression, anxiety, grief, existential suffering, or social suffering.

Before you read in Chap. 3 how many American adults live with extreme suffering, guess the percentage. Of course, it depends on how one defines 'extreme,' but reflects on the question before and after digesting the statistics.

Another important question is how much extreme suffering affects people's quality of life (QOL). The answer may surprise you. Finding so much suffering in a contemporary, affluent society raises the possibility that affluence itself, through lifestyles and beliefs produces types of suffering not typically found in poverty stricken nations.

Chapter 4 shifts to a global perspective and offers pioneering indicators for both subjective and objective suffering country by country. Besides ranking countries by their degree of suffering, the chapter notes how social support networks seem to help people living in different cultures cope with suffering more easily.

Alternative approaches to the alleviation of suffering depend upon the type of suffering, but all types need to be addressed on both the individual and institutional levels. Data comparing nations as well as states in Chap. 5 show the misalignment between suffering and available care resources that may help relieve those who suffer. A major finding is that global inequality is a major cause of suffering and widens gaps in care for those who suffer.

Working toward ending needless suffering is both a personal value and a public good that offers hope to those who suffer now or in the future. Chapter 6 reviews the ethical grounds for alleviating suffering. It also discusses strategies for relief of suffering and notes how the relief of suffering has to be both an individual and a collective effort. Recommendations are offered for incorporating the relief of suffering more fully into social policy for development as well as for individual decision-making.

Contents

Chapter 1
Conceptualizing Human Pain and Suffering

When we suffer, our quality of life declines—it is an intuitive idea. What is not so obvious, however, is that by intertwining suffering and quality of life in our *thinking*, we can better understand and cope with suffering (whether our own or others'). To begin pulling these concepts together, I highlight relevant social scientific literature and suggest eight frames or ways of thinking about and investigating human suffering. In later chapters, I will have examples from stories about suffering and statistics showing the spread of suffering, both national and global.

1.1 The Suffering Concept

Imagine yourself undergoing major surgery 200 years ago. While some cultures had used pain-relieving herbs and other natural substances for millennia, you are in Europe or America, where such palliatives are not yet used. You will not be given anesthesia (beyond, perhaps, some alcohol). Essentially, you are facing torture. This is what happened to Frances Burney, a wealthy English writer living in France in 1810. She left a vivid story of suffering as she described six surgeons' work to remove her breast tumor. Journal entries of her unimaginable pain have been described by Dormandy (2006) and preserved by Hemlow (1975). Her pain, unchecked by any anesthesia, sears the page:

> When the dreadful steel was plunged into my breast, I released an unremitting scream.... I felt the knife rackling against the breast bone, scraping it while I remained in torture.... When I opened my eyes I saw the good Dr. Larrey, pale nearly as myself, his face streaked with blood, and depicting grief, apprehension and almost horror (Hemlow 1975).

Burney's recollection is a monument to the raw pain and suffering of both patients and doctors. The doctor's suffering, resulting from compassion, attests to the reality of collective—or social—suffering.

For those of us living in an era of high-quality anesthetics and laser surgery, Burney's agony reads like primitive depictions of hell. Through her words,

R. E. Anderson, *Human Suffering and Quality of Life*, SpringerBriefs in Well-Being and Quality of Life Research, DOI: 10.1007/978-94-007-7669-2_1,

we empathize. Yet as you read this sentence, millions of people suffer in dark corners of the globe, just as millions have in the past, and, potentially, millions will in the future (Amato 1990). Every day, our fellow human beings face torture, rape, and excruciating trauma (Bourdieu et al. 2000; Dormandy 2006; Trachtenberg 2008; Vollman 2005).

Now, not all pain and suffering is extreme, bordering on the unbearable. Pain and suffering range from the infinitesimal to the unimaginably excruciating. And both pain and suffering may last seconds or lifetimes. They may be fleeting or chronic.

Pain and suffering may also be individual or social. Often we cut a finger, occasionally a friend dies, but such suffering is not distributed evenly across social strata, much less the globe (Anderson 2011, 2012; Bock 2011; Diener et al. 2009). While severe suffering from violence and injury occur more often in the Global South, particularly in pockets of poverty, studies in western societies generally conclude that at least 20 % of adults suffer from chronic pain, the reoccurrence of severe pain over several months or longer (Breivik et al. 2006; Chabal 2009; Collier 2007; Nagappan 2005; Kleinman 2009a; b; 2011). Suffering is pervasive, if not always shared.

In this book, the word 'suffering' will be used as an all-inclusive term, subsuming pain. However, Table 1.1, which identifies three categories of suffering and provides a brief entry of descriptors for each, categorizes pain as separate from other types of suffering. Our language is filled with words that imply affective or emotional responses to events or objects that result in negative feelings, many of which are listed in Table 1.1. For example, grief as a type of suffering is viewed by Charmaz and Jilligan (2006) as a composite of many emotions and cognitions including fear and sorrow.

In the spirit of Cassel (2004) and Chapman and Volinn (2005), who defined suffering as perceived threat or damage to a sense of self, here suffering is defined as distress resulting from threat or damage to one's body or self-identity. Suffering can vary in intensity, duration, awareness and source. Physical suffering is the subset of distress resulting from threat or damage to one's physical being, whereas mental suffering is distress perceived as originating in one's cognitive or affective self-identity. Self-identity is the set of characteristics and their meanings observed when one looks at oneself.

Physical suffering is equated with pain, even though it often co-occurs with mental suffering (Black 2005; Carr et al. 2005; Livingston 1998; Morris 2002; Wilson et al. 2009), while mental suffering includes cognitive suffering (thoughts that produce suffering) and emotional suffering (Francis 2006).

For present purposes, social suffering is defined as suffering whose sources are social collectivities and/or social institutions. Social suffering, which will be discussed at much greater length in the next section, differs in that it refers to the social contexts that shape the suffering of both individuals and collectivities. Social suffering typically co-occurs with other types of suffering, results from social forces, and results in social change (Das et al. 2001; Farmer 1997; Kleinman 1988;

Table 1.1 Words associated with common types of suffering

Suffering type	Words for suffering
Physical suffering (Pain)	Agony, discomfort, excruciation, hurt, incapacitation, torture, torment, soreness, acute pain, chronic pain, extreme pain, excruciating pain, unimaginable pain
Mental suffering	Anguish, angst, anxiety, addiction, distress, troubled, craving, post-traumatic stress disorder, compulsive disorder, loss, mourning, grief, sadness, disgust, irritation, anger, rage, hate, contempt, jealousy, envy, frustration, heartbreak, fear, panic, horror, indignation, shame, guilt, remorse, regret, resentment, repentance, embarrassment, humiliation, boredom, apathy, confusion, disappointment, hopelessness, doubt, emptiness, homesickness, loneliness, rejection, pity, self-pity, nervousness, restlessness, minor depression, chronic depression, severe depression, hopelessness, self-worthlessness, spiritual confusion, purposelessness, other types of loss of meaning
Social suffering	Social exclusion, discrimination, ostracized, persecution, incapacitation, disability, shame (self-ostracized), distrust, relative deprivation, subjugation, atrocity, homelessness, unemployment, social rejection, discrimination, bullied, disability, blindness, deafness, bedridden, hunger, war, civil violence, survival risk factors

Kleinman et al. 1997; Nordgren et al. 2011; Wilkinson 2005a). Genocide, battlefield slaughters, and lynching are well-known examples. Research on social suffering has uncovered that those affected by such dreadful events suffer in part from a devastating loss of their identity as human beings (Bourdieu et al. 2000; Kleinman et al. 1997; Wilkinson 2005a, b).

Existential suffering (later combined with mental suffering) is the result of struggles with the meaning of one's existence (Langle 2008). This may seem like a lofty idea, but you might think of it as a struggle in which you question the meaning of your life (or life itself). A common course of existential suffering is confrontation with death and other threats to one's existence. Williams (2004) interviewed low-income cancer victims receiving end of life care. In many instances, the patients' suffering was compounded by wondering how their impending death could square with their beliefs about life's meaning. On top of that, some felt left out or treated as non-persons as death approached. Here is how a 42-year-old man described the experience:

> People talk as if you're not there. One of mother's friends died of cancer last week, and people around me were talking all about the funeral, like they didn't even think it might bother me. It gets to me and makes me feel my life isn't worth anything compared to theirs (Williams 2004).

This narrative demonstrates how social and existential suffering may occur together, amplifying the degree of tragedy and suffering.

1.2 A Taxonomy for Pain and Suffering

Pain is such a complex phenomenon that thousands of scientists have yet to isolate and understand all its aspects. A noted figure in the science of pain, Livingston (1998), said at one point that "nothing can be properly called pain unless it is consciously perceived as such." The word 'pain' is derived from the Greek *poine* and the Latin *poena*, both of which referred primarily to punishment or penalty. (Like other primitive peoples, the early Greeks believed their many gods handed out rewards and punishments and both were generally received by people as pain.)

Aristotle spoke of pain and pleasure as "passions of the soul," and claimed "wherever there is sensation there is also pain or pleasure" (Livingston 1998). To this day, many associate pain and suffering with matters of the soul and spirituality. Because severe pain easily preoccupies the mind, we should not be surprised that people often seek to know why and how they became the victims of the pain. Suffering may lead to speculation on existential matters like the meaning of pain and suffering in one's life and in the larger schemes or purposes of life. In discussing the sociology of emotions, Francis (2006) asserts that emotions play a major role in pain and suffering, how the victim interprets the meaning of pain affects the emotions evoked.

In a modern definition that takes into account emotion (if not spirituality), the International Association for the Study of Pain states: "Pain is an unpleasant sensory and emotional experience associated with actual or potential tissue damage" (Merskey and Bogsuk 1994). Thus, pain is perceived as a negative emotional *and* sensory experience (Brattberg et al. 1996; Das 1997b; Sontag 2003).

Pain and suffering are often used synonymously, but suffering also refers to psychological or social hurt (whether or not that hurt originated from the negative physical sensation we call pain). Suffering also encompasses social affliction and stress, as well as the emotional component of pain. As pain has a physiological and/or neurological character, and because it affects health so directly, it has been extensively investigated by scientists. Furthermore, pain management has become a major health care field and a large industry in western societies. The phrase 'pain and suffering' even has a special meaning within the legal system.

To consider suffering separately for a moment, though, we can start with its origins. The word suffering emerged from Middle English word *suffrir* and the Latin word *suffero*, both of which were defined as being 'long-suffering' or facing a burden of pain with patience. Over several hundred years, the word lost its reflection of endurance and remained a verbal representation of hardship, distress, and turmoil.

Suffering is used in so many different ways that the very word might become a barrier rather than an aid to understanding. Carefully considered taxonomies can prevent confusion, and that is why I explored the many synonyms and meanings of suffering in Table 1.1. To refine our thinking, I then constructed Table 1.2 by reviewing the use of words like suffering, pain, misery, and distress in the academic literature and in popular works. I compared the best sources to see where

Table 1.2 Examples of suffering by their source (rows) and type (columns)

	Physical suffering (Pain)	Mental suffering	Social suffering
Primarily produced internally	Suicide, self-flagellation, self-abuse	Paranoia, low self-regard, unwarranted fear, anger, other negative emotions	Worry, fear, perceived violence, imagined threats such as bullying, shame, envy, greed, jealousy
Primarily caused externally by nature	Injuries from natural disasters	Loss of goods and others living beings from natural disasters	Collective starvation, homelessness, injury
Primarily caused externally by persons or small groups	Torture, rape, bullying	Threats of hostile act (e.g., rape, bullying)	Injuries, rape, gang violence and assaults, stalking
Primarily caused externally by institutions, societies, social forces	Hunger, starvation, illness, injury, poverty, sexism, political violence	Threats of persecution (e.g., racism, PTSD)	War casualties, collective rape, incarcerations, discrimination, stigma

they agreed on the dimensions or domains of suffering. These dimensions include not only different types of suffering, but diverse causes, outcomes, and meanings.

In the interest of brevity, only three categories of suffering appear as columns in Table 1.2. Here, existential suffering is considered mental suffering.

You will notice that each category represented by a row in the table begins with the word *primarily*. Very often, any given instance of suffering has multiple causes and multiple processes as represented by the columns of the table.

Now, human suffering can only be fully understood from the accumulation of knowledge about its causes, contexts, and results. Suffering is *so* broad that knowledge is needed from many disciplines, including the humanities, social sciences, biological sciences, and professional health care. However, discerning the mutual interplay between suffering and the quality of life depends largely upon knowledge and tools from within social science research. This is why qualitative, quantitative, historical, and comparative methods provide the basis for my investigation in the rest of the chapter.

Social suffering is a relatively new label. The term emerged from anthropologists and sociologists studying pain and suffering ethnographically. Medical anthropologists Kleinman et al. (1997) wrote the first book titled *Social Suffering*, and they continue to work on the construct, most recently applying it to global humanitarian policy (Kleinman 2010b, 2011; Farmer 2005, 2006). Das (1997b) and another medical anthropologist, Morris (2002), have helped explicate the concept. Sociologist Wilkinson (2005a, b) devoted his book on suffering to applying and enhancing the notion of social suffering, which he defined as suffering produced by social forces, rendering the victim without a sense of being human and worthwhile. Using the Holocaust, the Rwandan Genocide, and similar atrocities, he and the other investigators of social suffering repeatedly emphasized how large-scale events leave their victims feeling like their humanity is superfluous.

Kleinman (2010a) also defined social suffering as the suffering caused by social forces, but emphasized social institutions, global systems, and culture as the culprits. Kleinman argues that the concept is meant to mix together social and health problems of every sort. Scholars in this line of thinking focus on 'lived experience', the ranges of harms done to the victims of suffering, and the need for a radical reappraisal of contemporary moral and political values. Wilkinson (2005a, b) has written that the aim of a framework of social suffering is to reflect a moral demand to reinterpret the meaning of modern history, to 'humanize' the ways we all relate as global citizens. Perhaps the greatest merit of the concept of social suffering is that it points out not only how horrifyingly inhuman many global acts continue to be, but also the role that institutional policies may play in producing greater suffering, even though the policies have been intended to relieve suffering.

Here is an example of social suffering from a story on the website Reasons to Go On Living (thereasons.ca). The author was brought up as a strict Catholic. At the age of 16, she discovered that she was in love with her best girlfriend.

> After a year of struggling with my religious beliefs, I felt like there was no way out for me…. I was driving and came within seconds of stopping my car on railroad tracks and committing suicide. Looking back 20 years later, it shocks and angers me that homophobia and heterosexism almost killed me.

It is useful to distinguish collective suffering from those instances of suffering from the suffering produced by social forces. Thus, Table 1.2 distinguishes social suffering from nonsocial types of suffering (with columns) and distinguishes institutional causes of suffering from individual and small group sources of suffering (with rows). (Both the bottom row and the right-hand column represent social suffering.)

That these different types of suffering can be distinguished does not mean that they do not overlap or co-occur. Note, for instance, that rape is listed in several different cells of Table 1.2; the suffering resulting from such violence can be both individual and social. A victim's suffering can also be a consequence of both individual and societal forces.

1.3 Frames for Thinking about Suffering

To better understand the role of suffering in history (as well as in present day global society), it is helpful to identify and trace the major frames (or points of view) that people use to organize their thoughts about it. Frames are complex perspectives that structure thought and build a rationale for a particular rhetoric, ideology, ethical principle, or social movement. Frame analysis explores whether the frame may foster social change.

Table 1.3 Eight frames for suffering by human centeredness and focus

	Individual focused frames	Collectivity focused frames
Supra- human centered	(1a) Suffering as punishment *(It is the price you pay for doing what you want to.)* (1b) Suffering as reward *(It is how you learn what not to do.)*	(3a) Suffering as natural destiny *(You are encountering destiny.)* (3b) Suffering as manageable *(Science can make suffering bearable.)*
Human centered	(2a) Suffering as craving *(You can have joy from victory over craving.)* (2b) Suffering from altruistic action *(Your loss will benefit others.)*	(4a) Relief of suffering as human purpose *(Relief of suffering gives you purpose.)* (4b) Relief of social suffering as progress in quality of life *(Relief gives us progress.)*

In the table below, I have identified eight frames from a review of the commonalities and differences in the literature on the meaning of pain and suffering. Special attention was given to the major scholars of pain and suffering such as Dormandy (2006); Cassell (2004); Morris (2002); Nordgren et al. (2011), and Wilkinson (2005a, b). Table 1.3 shows how the eight frames fall across two different dimensions: human versus supra-human centeredness and individual versus collectivity focus. It is important to note that this is the first time these frames have been outlined as related to the meaning of suffering. It is also novel in that my scheme casts altruism as suffering and includes progress in quality of life as explicitly linked to the meaning of suffering.

Within each cell, two frames are listed, and for each of these frames, an example statement is given in italics. The top row represents frames for suffering that center on supra-human phenomenon, namely the supernatural and nature or destiny. The bottom row, in contrast, features human centered frames, in that human beings can each choose to interpret the suffering for their own purposes.

The assignment of suffering frames to a given cell is not absolute but heuristic (illustrative of the implications of the frames), and the implications of each frame may change over time. For example, the frames in cell 1 were dominant in pre-modern times, but they have been modified and adapted to contemporary, secular culture.

1.4 Suffering as Punishment

The first frame, *suffering as punishment*, was predominant from the earliest historical periods down through the middle ages. During both the era of animist religions and the early era of organized religions (including Judaism, Christianity, and Islam), pain and suffering were attributed to higher powers (Bowker 1970; Dormandy 2006; Kruse and Bastida 2009). God or the gods were thought to determine when, where, how, and what suffering was distributed among human beings, and this punishment was doled out as an indication of the higher powers' displeasure with humans' attitudes and behaviors. As already noted, 'suffer' once implied long-suffering or patience—necessary to cope with the severe and sometimes arbitrary suffering of everyday life. By aligning their behavior with what they saw as the will of God or the gods, people believed that they were maximizing their relief from suffering. Of course, there will still be many people today who frame suffering primarily as punishment.

1.5 Suffering as Reward

The second frame, *suffering as reward*, first emerged from the punishment frame. Since suffering was interpreted as a sign of displeasure from the supernatural, it was also seen as a reward. A divine power was indicating which behaviors were off-limits, which meant you could avoid future suffering by avoiding the behavior that brought on your suffering. Some religious groups have even presumed that, because we can learn from suffering, it is a desirable, laudable condition that should be exalted (Ashwell 2011; Beke 2011; Ghadinian 2012). In the thirteenth century, a group of Roman Catholics, known as the Flagellants, took this practice to its extreme ends, marching through the streets whipping themselves. After several deaths, the Church officially withdrew its approval of these events (Bean 2000). Still, some contemporary religions will celebrate holy days devoted to suffering. Adherents, too, believe that withstanding pain is a holy act, so using medications or other sources of relief is less desirable than fully experiencing suffering. Author and Trappist monk Thomas Merton (1955) said, "We must see suffering not as a destructive power but as a transcendent gift from the Divine."

Ironically, we could even see the exalting of suffering in the 2012 presidential campaign in the United States. During a Republican primary forum, four candidates took turns telling their story of extreme suffering and how it had made them a better Christian and closer to God. One candidate even said, "Suffering… is not a bad thing, it is an essential thing in life" (Jacoby 2011). Unfortunately, this belief in suffering as a good leads many to take a stand against government funding for the poor and others who suffer.

New institutions in western legal systems also indirectly support the frame that suffering is a reward. In the United States, tort cases in which people seek compensation for pain and suffering tend to result in considerable economic payoff (Rodgers 1993). Conventional norms in the legal and insurance systems for different types of suffering even provide guidelines for the economic payment due families for the death of a family member. Logically, the idea is that victims did not bring their suffering upon themselves, and so someone responsible should bear the 'punishment' in the form of a financial payment or other settlement.

1.6 Suffering as Craving

The frame for *suffering as craving* is a very popular attitude toward suffering in eastern cultures. Equivalent notions of 'addiction as suffering' and 'unrestrained pleasure' as suffering are common in most religious traditions.

The following quote is attributed to Socrates: "If you don't get what you want, you suffer; if you get what you don't want, you suffer; even when you get exactly what you want, you still suffer because you can't hold on to it forever" (Millman 2006). Millman gives this notion a western psychological slant with "Pain is objective and physical; suffering is our psychological resistance to such events" (2006). As noted by Hurst (2011), Merton (1961) taught "contemplation as a way of living in awareness, allowing us to integrate suffering into life." Aristotle advocated a middle way between excess and asceticism, not unlike Buddha's middle path (Shields 2012).

Buddhism directly teaches that 'Pain is inevitable; suffering is optional,' and 'the origin of suffering is craving.' Craving is interpreted by some as egocentric habits of mind (Targ and Hurtak 2006). The Buddha warned that all pleasurable sensations lead to craving and craving can take root (Dalai Lama 2011; Dalai Lama and Goleman 2003). Attaching to that craving causes suffering (as with addiction). Thus, the Buddha advocated the Middle Path, which avoids the extremes of a life of unrestrained pleasure-seeking *and* a life of extreme denial and suffering (Nikaya 1971). Buddhist practice consists of learning to live without specific pleasures by engaging in mindfulness and loving kindness for all living beings. Mindfulness is a meditative practice intended to keep the mind from its tendency to cling to emotions such as anger and hatred and to entertain thoughts of retribution and self-pity (Siegel 2010). As a Buddhist takes up this life of mindfulness and contemplative practice, cravings are less able to take root (Bernhard 2010).

Another metaphor for this process is uniting with a greater universal consciousness. Other religions try to define rules or standards for people to balance pleasure with indulgence such that addictive craving is avoidable. Few are effective, though, because anger, greed, over-indulgence, and other types of suffering that result from craving are commonplace, if not rampant, in most societies (Pruett 1987).

1.7 Suffering from Altruistic Action

This frame, *suffering from altruistic action*, is not intended to suggest that all altruism results in suffering. Instead, this perspective merely points out that sometimes an altruistic action requires a sacrifice on the part of the compassionate giver (Corbett and Fikkert 2012). Altruistic action is regard for another as an end in itself. It involves providing assistance without consideration (at least at the time) of whether you might benefit or receive a reward. Sometimes this is called empathic-altruism, because without empathy, the other's welfare alone will not be seen as a legitimate goal. The most important implication of this type of altruism is that it may put one at risk—possibly of suffering.

In the well-known Biblical story of the Good Samaritan, the man from Samaria gave freely of his time and resources to the injured victim. It is not known whether the Samaritan suffered serious loss. Elsewhere, Jesus tells his followers, "If you will be perfect, go and sell what you have, and give to the poor, and you shall have treasure in heaven: and come and follow me." This admonition implied anyone with material wealth might risk suffering. However, the implied promise was that anyone suffering from the loss of material goods due to altruistic action would be compensated by spiritual or moral rewards.

At the heart of the frame of altruistic suffering is the concept of sacrifice or self-sacrifice (Das 1997a). The potential suffering resulting from empathic-altruism that requires significant sacrifice is often considered too high a trade-off, even by devout Christians. The overall reduction of suffering in the world is probably not possible, though, without the willingness of many to altruistically accept personal risk for the benefit of the common good (Schopenhauer 2004). Sacrifice is not limited to material goods with explicit economic value, but it may include more elusive but highly valued social goods like important interpersonal relationships.

1.8 Suffering as Natural Destiny

Hundreds of thousands of people die and many more suffer severely every year from natural disasters. That every human will die and leave behind untold grief is a fact of life. In modern societies, most people believe in and accept seemingly random suffering—it is seen as natural destiny (Cassell 2004; Ferrell and Coyle 2008). We still tend to assess whether we (or other human beings) might have caused the suffering. Often we don't have sufficient knowledge to make a precise determination. Human beings play an indirect role in producing suffering by such actions as deforestation and air pollution. Many other natural forces determine the specific calamities and victims of the resulting suffering.

Decay, death, and suffering are so common across the universe that some hold the position that suffering is an inherent and necessary aspect of how the universe works. Wertenbaker (2011) takes the side of those who view suffering as an inevitable outcome of the separation and isolation of individual cells, organs, persons, and planets because they have lost their natural unity with a single universe or consciousness. Yet, it is hard to envision any kind of progress without some separation among these building blocks of the universe (Upton 2011).

Extreme views of suffering as a dominating principle of the universe generate more pessimism than seems warranted. If the universe thrives on pain and suffering, the goal of ending preventable suffering seems idealistic at best. In the absence of knowledge that suffering and devastation will inevitably lead to destruction of life, wisdom would lead human beings to pursue initiatives that eradicate all humanly-preventable suffering in the interest of the preservation of the species.

1.9 Suffering as Manageable

In a perfect world, could we eradicate suffering? Contemporary thought has evolved from historical frames to utopian visions in which advancing technology, especially pain medication, can end suffering. Indeed, in the past century, pain control has revolutionized healthcare and millions' quality of life. Were it not for the rapidly rising life expectancy in most countries, it would be possible to claim a rapid decline in pain and suffering; because of expanded lifespans, however, the average person will naturally experience more illness and injuries. Also, the populations in countries *lacking* the luxury of pain medications continue to rise at the fastest rates.

Pinker (2011) compiled a highly compelling case that violence has in general been on the decline throughout human history. He makes his case using rates of a wide variety of social indicators. A major flaw in applying his conclusions to suffering is that he does not take population growth into account, which less increasing longevity, both of which exacerbate the rise in suffering. An exhaustive study remains to be done regarding the historical rise and fall of suffering. Meanwhile it is clear that contemporary political decisions fail to weigh in on suffering. For example, the war of terror, a response to 3,000 killed in by the 9/11 attack, has already produced ten times as many extreme sufferers in the Middle East, including deaths, displaced families, extremely painful injuries and others dedicated now to bombing and maiming their enemies.

It is difficult to disparage the pain management movement except in so far as it neglects the humanity of those served. Given widespread inequality, not all who need pain management can get it. Many in western countries cannot afford expensive pain medications, to say nothing of the billions in the developing world who lack access. Even those served with pain relieving drugs are not always *well* served; they can be treated as objects rather than as individuals with unique needs and concerns. Finally, the fear that the sufferer may become addicted to the drugs sometimes results in the withholding of pain treatments. Where physicians have been prosecuted for being too liberal in dispensing drugs, other doctors may become over-cautious or even stingy. Large subpopulations of those who need the pain relief do not receive it. Perhaps even more insidious is a doctor's failure to provide pain relief, based in a non-medical belief that suffering is useful punishment for those who appear to have been careless, for example, an alcoholic with liver disease or a person who was injured committing a crime (Melzack 1990; Taylor 2007).

1.10 Relief of Suffering as Human Purpose

The principle purpose of many (if not most) humans is self-promotion. They hope to obtain (or maintain) comfort, power, popularity, and wealth. Some, though, are driven primarily by a feeling of moral responsibility for others' wellbeing (Kleinman and van der Geest 2009; Mayerfeld 2005; Tronto 1993; Williams 2008). The most common literary symbol of such a commitment to others is the Christian Good

Samaritan—people with humanitarian commitments to helping others, no matter their race or stature, are sometimes called good Samaritans. A similar sentiment motivates a recent campaign to get hundreds of thousands of people (regardless of faith) to commit themselves to the Charter for Compassion (Armstrong 2011).

For those whose purpose is love, compassion, or helping others, suffering provides a basis by which to prioritize limited time and attention (Johnson and Schollar-Jaquish 2007). Helping those who suffer more is generally seen as more fulfilling. Further, since the traditional definition of compassion is a desire to relieve another's suffering, this work becomes the yardstick by which to measure an authentic life; suffering is an indirect source of meaning in the Samaritan's life. Contributing to humanity in this sense could mean helping a few close friends or all seven billion people alive today.

The mission to relieve suffering does not require one-to-one contact. It can be accomplished by providing time and resources to global relief organizations. By giving to varied causes or helping a variety of different types of people in need, you increase the likelihood that your pro-social actions will have benefited a person or several people. While positive feedback is not mandatory for gaining purpose and satisfaction from compassionate actions, it does help prop up and support the energy put into reducing the suffering of others.

1.11 Relief of Social Suffering as Progress in Quality of Life

The process of meaningful relief of others' suffering, as discussed in the preceding section, applies to this frame as well. When you are relieving another's suffering, you are also improving their quality of life. This frame is uniquely justified by its emphasis on quality of life as a concrete human need and its emphasis on *social* suffering as a qualitatively different type of suffering.

As a common phrase, 'quality of life' (QOL) goes back only a few decades. However, in the twenty-first century, the concept has become rather popular, especially within research on health and economics (Land et al. 2012; Mukkerjee 1989). There is even a professional group called the International Society for Quality of Life Studies, and it publishes several journals with 'quality of life' in their titles. Many national and international policy reports also use the phrase, sometimes equating it with general well-being and/or happiness (Jordan 2012). The governments of several nations are now using the concept in attempting to construct new measures of national or human progress.

1.12 Conclusions

At the beginning of this chapter, I began our exploration with the assumption that suffering and quality of life are intimately connected. Major suffering undermines the quality of life, which is also called flourishing, thriving, or well-being.

In fact, suffering is so intertwined with quality of life that it may be useful to treat suffering as an indicator of negative quality of life (Eckermann 2012). It also is useful to conceptualize suffering as both a *component* and an *outcome* of quality of life.

If suffering and its relief are viewed as pertaining only to oneself, then so is quality of life. But if we are only concerned about the suffering of *others*, then their quality of life would be the focus. The implication of this is that in measuring subjective quality of life, it may be necessary to distinguish a person's perception of their own quality of life from that of others for whom the person feels concern and wishes to relieve suffering. Conceptualizing quality of life separately for oneself and others of importance could enhance not only our understanding of quality of life, but the mapping the social circles of concern to different types of individuals. The outcome would be greater understanding of differences in the meanings of suffering.

Social suffering, as defined by Wilkinson (2005a, b) and others writing on the concept, suggests that social suffering deserves high priority by both social scientists and policy makers. Wilkinson argues that the idea of social suffering could lead to a reinterpretation of the meaning of modern history, humanizing the ways we relate to one another as global citizens.

At a minimum, the concept is likely to lead to a fuller acknowledgement of what happens to the humanity of those who suffer, whether under the extremes of economic hardship, social injustice, or political oppression. In dwelling on what suffering *does* and developing new ways of thinking about the pain and distress of embodied experience, we may begin to contribute more substantially to *global* quality of life. The study of social suffering may, if we are to think and hope broadly, will increase the effectiveness of the next generations of scholars, activists, and other humanists in fostering the moral and political regeneration of the forces for common good.

References

Amato, J. A. (1990). *Victims and values—a history and a theory of suffering*. New York: Greenwood Press.

Anderson, R. E. (2011). *World suffering—Conceptualization, measurement, and findings*. Paper presented at the 2011 meeting of the American Association for Public Opinion Research (AAPOR) in Phoenix, AR. Retrieved Feb 17, 2013, from http://tinyurl.com/77b8d4u.

Anderson, R. E. (2012). *Human suffering and the measurement of human progress*. Paper presented at the 2012 Forum of the International Sociological Association in Buenos Aires, Argentina. Aug 1, 2012.

Armstrong, K. (2011). *Twelve steps to a compassionate life*. New York: Anchor.

Ashwell, R. (2011). Suffering and the way. *Parabola*, 34(1, Spring), 58–61.

Bean, J. (2000). *Flogging*. Emeryville, CA: Greenery Press.

Beke, J. (2011). Intentional suffering. *Parabola*, 34(1, Spring), 68–71.

Bernhard, T. (2010). *How to be sick—A Buddhist-inspired guide for the chronically ill and their caregivers*. Boston, MA: Wisdom Publications.

Black, H. K. (2005). *Soul pain: The meaning of suffering in later life*. Amityville, NY: Baywood Publishing Company, Inc.

Bock, J. (2011). Humanitarian aid and the struggle for peace and justice: Organizational innovation after a blind date. *Journal of Sociology and Social Welfare, 38*(2), 37–51.
Bourdieu, P., et al. (2000). Understanding. In P. Bourdieu, et al. (Eds.), *The weight of the world: Social suffering in contemporary society*. Stanford, CA: Stanford University Press.
Bowker, J. (1970). *Problems of suffering in religions of the world*. Cambridge, UK: Cambridge University Press.
Brattberg, G., Parker, M., & Thorslund, M. (1996). The prevalence of pain among the oldest old in Sweden. *Pain, 67*(1), 29–34.
Breivik, H., Collett, B., Ventafridda, V., Cohen, R., & Gallacher, D. (2006). Survey of chronic pain in Europe: Prevalence, impact on daily life, and treatment. *European Journal of Pain, 20*, 287–333.
Carr, D. B., Loeser, J. D., & Morris, D. B. (Eds.). (2005). *Narrative, pain and suffering (Progress in pain research and management*. Seattle, WA: International Association for the Study of Pain Press.
Cassell, E. J. (2004). *The nature of suffering and the goals of medicine*. Oxford, UK: Oxford University Press.
Chabal, P. (2009). *Africa: The politics of suffering and smiling*. London: Zed Books.
Chapman, C. R., & Volinn, E. (2005). Narrative as a window on chronic disabling back pain. In D. B. Carr, J. Loeser, & D. B. Morris (Eds.), *Narrative, Pain, and Suffering* (pp. 73–86). Seattle, WA: International Association for the Study of Pain Press.
Charmaz, K., & Jilligan, M. J. (2006). Grief. In J. E. Stets & J. H. Turner (Eds.), *Handbook of the sociology of emotions* (pp. 519–543). New York: Springer.
Collier, P. (2007). *The bottom billion—Why the poorest countries are failing, and what can be done about it*. New York: Oxford University Press.
Corbett, S., & Fikkert, B. (2012). *When helping hurts—How to alleviate poverty without hurting the poor and yourself*. Chicago, IL: Moody Publishers.
Dalai Lama. (2011). To be free of suffering. *Parabola*, 36(1, Spring), 26–30.
Dalai Lama., Goleman, D. (2003). *Destructive emotions—How can we overcome them?* New York: Bantam Books.
Das, V. (1997a). Sufferings, theodicies, disciplinary practices, appropriations. *International Journal of Social Science, 49*, 563–567.
Das, V. (1997b). Language and body: Transactions in the construction of pain, In A. Kleinman, V. Das, & M. Lock (Eds.), *Social Suffering*. Berkeley: University of California Press.
Das, V., Kleinman, A., Ramphele, M., Lock, M., & Reynolds, P. (Eds.). (2001). *Remaking a world: Violence, social suffering & recovery*. Berkeley: University of California Press.
Diener, E., Kahneman, D., Tov, W., & Arora, R. (2009). Income, aging, health and wellbeing around the world: Evidence from the Gallup World Poll. In E. Diener (Ed.), *Assessing well-being* (pp. 233–246). Oxford, UK: Springer.
Dormandy, T. (2006). *The worst of evils—The fight against pain*. New Haven, CT: Yale University Press.
Eckermann, E. (2012). The quality of life of adults. In K. Land et al. (Eds.), *Handbook of social indicators and quality of life research* (pp. 373–380). New York: Springer.
Farmer, P. (1997). On suffering and structural violence: A view from below. In A. Kleinman, V. Das, & M. Lock (Eds.), *Social suffering*. Berkeley: University of California Press.
Farmer, P. (2005). *Pathologies of power: Health, human rights and the new war on the poor*. Berkeley: University of California Press.
Farmer, P. (2006). Never again? Reflections on human values and human rights. In G. B. Peterson (Ed.), *The tanner lectures on human values*. Salt Lake City: University of Utah Press.
Ferrell, B. R., & Coyle, N. (2008). *The nature of suffering and the goals of nursing*. Oxford, UK: Oxford University Press.
Francis, L. E. (2006). Emotions and health. In W. E. Stets & J. H. Turner (Eds.), *Handbook of the sociology of emotions* (pp. 591–610). New York: Springer.
Ghadirian, A. M. (2012). *Creative dimensions of suffering*. Minneapolis, MN: BookMobile.

Hemlow, J. (Ed.). (1975). *The journals and letters of Fanny Burney*. Oxford, UK: Oxford University Press.

Hurst, V. (2011). The flight from disunity: Thomas Merton on suffering. *Parabola,* 34(1, Spring), 34–39.

Jacoby, S. (2011). Christian politicians exalt suffering in GOP campaign. *The Washington Post* (23 November). Retrieved Jan 27, 2013, from http://www.washingtonpost.com/blogs/spirited-atheist/post/christian-politicians-exalt-suffering-in-gop-campaign/2011/11/22/gIQAa9HVoN_blog.html.

Johnson, N., & Schollar-Jaquish, A. (Eds.). (2007). *Meaning in suffering—Caring practices in the health professions*. Madison, WI: University of Wisconsin Press.

Jordan, T. E. (2012). *Quality of life and mortality among children*. New York: SpringerBriefs.

Kleinman, A. (1988). *The illness narratives: Suffering, healing and the human condition*. New York: Basic Books.

Kleinman, A. (2009b). Unpacking global health: A critical sociology of knowledge III. Slide presentation. Retrieved Jan 20, 2013, from http://www.scribd.com/doc/90989880/Lecture-2-Unpacking-Global-Health-II.

Kleinman, A. (2010a). Caregiving: Its role in medicine and society in America and China. *Aging International, 35*(2), 96–108.

Kleinman, A. (2010b). Four social theories for global health. *The Lancet, 375*(9725), 1518–1519.

Kleinman, A. (2011). Anthropology and cross-cultural mental health: The major questions for future research in global mental health. (Slide presentation). Retrieved Jan 20, 2013, from http://www.segemi.de/files/kleinman_anthropology_cross-cultural_mental_health_6-2011_1_.pdf.

Kleinman, A., & van der Geest, S. (2009). 'Care' in health care: Remaking the moral world of medicine. *Medische Antropologie, 21*(1), 159–168.

Kleinman, A., Das, V., & Lock, M. (Eds.). (1997). *Social suffering*. Berkeley: University of California Press.

Kruse, N., & Bastida, E. (2009). Religion, suffering, and health among older Mexican Americans. *Journal of Aging Studies, 23*(2), 114–123.

Land, K. C., Michalos, A. C., & Sirgy, M. J. (2012). The development and evolution of research on social indicators and quality of life. In K. Land, et al. (Eds.), *Handbook of social indicators and quality of life research* (pp. 1–22). New York: Springer.

Langle, A. (2008). Suffering—an existential challenge: Understanding, dealing and copy with suffering from an existential-analytic perspective. *International Journal of Existential Psychology & Psychotherapy*, 2(1). Retrieved Jan 28, 2013, from http://journal.existentialpsychology.org/index.php?journal=ExPsy&page=article&op=view&path%5B%5D=115&path%5B%5D=58.

Livingston, W. K. (1998). *Pain and suffering*. Seattle, WA: International Association for the Study of Pain Press.

Mayerfeld, J. (2005). *Suffering and moral responsibility*. NY: Oxford University Press.

Melzack, R. (1990). The tragedy of needless pain. *Scientific American*, 262(2). Retrieved Jan 28, 2013, from http://druglibrary.org/schaffer/asap/sa1.htm.

Merskey, H & Bogsuk N, (eds.) (1994). *Classification of chronic pain* (2nd ed.) (IASP Task Force on Taxonomy). Seattle, WA: International Association for the Study of Pain Press. (Also available online at www.iasp-pain.org).

Merton, T. (1955). *No man is an island*. New York: Harcourt, Brace Jovanovich.

Merton, T. (1961). *New seeds of contemplation*. New York: New Directions.

Millman, D. (2006). *The journeys of socrates: An adventure*. New York: HarperOne.

Morris, D. B. (2002). *The culture of pain*. Berkley: University of California Press.

Mukherjee, R. (1989). *The quality of life—Valuation in social research*. Newbury Park, CA: Sage.

Nagappan, R. (2005). *Speaking havoc: Social suffering & South Asian narratives*. Seattle: University of Washington Press.

Nikaya, S. (1971). *The word of the Buddha*. Kandy, Sri Lanka: Buddhist Publication Society.

Nordgren, L. F., Banas, K., & MacDonald, G. (2011). Empathy gaps for social pain: Why people underestimate the pain of social suffering. *Journal of Personality and Social Psychology, 100*(1), 120–128.

Pinker, S. (2011). *The better angels of our nature: Why violence has declined.* NY: Penguin Books.
Pruett, G. E. (1987). *The meaning and end of suffering for Freud and the Buddhist tradition.* Lanham, MD: University Press of America.
Rodgers, G. B. (1993). Estimating jury compensation for pain and suffering in product liability cases involving nonfatal personal injury. *Journal of Forensic Economics*, 6(3), 251–262.
Schopenhauer, A. (2004). *On the suffering of the world.* London: Penguin Great Ideas.
Shields, C. (2012). Aristotle. In N. Z. Edward (Ed.), *The stanford encyclopedia of philosophy.* Retrieved May 13, 2013, from http://plato.stanford.edu/archives/sum2012/entries/aristotle/.
Siegel, R. D. (2010). *The mindfulness solution.* New York: The Guilford Press.
Sontag, S. (2003). *Regarding the pain of others.* New York: Picador.
Targ, R., & Hurtak, J. J. (2006). *The end of suffering—Fearless living in troubled times.* Charlottesville, VA: Hampton Roads Publishing.
Taylor, A. L. (2007). Addressing the global tragedy of needless pain: Rethinking the United Nations Single Convention on Narcotic Drugs. *Journal of Law, Medicine & Ethics, 35*(4), 556–570.
Trachtenberg, P. (2008). *The book of calamities: Five questions about suffering and its meaning.* NY: NY. Little, Brown and Company.
Tronto, J. C. (1993). *Moral boundaries: A political argument for an ethic of care.* New York: Routledge.
Upton, C. (2011). Metaphysics of suffering. *Parabola,* 34(1, Spring), 72–83.
Vollman, W. T. (2005). *Poor people.* New York: HarperCollins.
Wertenbaker, C. (2011). The cosmic necessity of suffering. *Parabola*, 34(1, Spring), 44–51.
Wilkinson, I. (2005a). *Suffering: A sociological introduction.* Indianapolis, IN: Polity.
Wilkinson, I. (2005b). From the sociology of risk to a critical sociology of suffering. Presentation at the SCARR Network Conference. Retrieved March 9, 2013, from http://www.kent.ac.uk/scarr/events/finalpapers/wilkinson.pdf.
Williams, B. R. (2004). Dying young, dying poor: A sociological examination of existential suffering among low-socioeconomic status patients. *Journal of Palliative Medicine*, 7(1), 27–37.
Williams, C. R. (2008). Compassion, suffering and the self: A moral psychology of social justice. *Current Sociology, 56*(5), 5–24.
Wilson, K. G., Chochinov, H. M., Allard, P., Chary, S., Gagnon, P. R., Macmillan, K., et al. (2009). Prevalence and correlates of pain in the Canadian National Palliative Care Survey. *Pain Research & Management, 14*(5), 366–379.

Chapter 2
Narrative Accounts of the Agony of Suffering

2.1 The Significance of Narratives of Pain and Suffering

A narrative is a story that has a teller, a listener, a language, characters, plot, and the dimension of time. Narratives and suffering have a long, common history; in fact, healers, doctors, and other health care workers have little else on which to base diagnoses of pain or suffering than their patients' words. A personal story often yields important details, which is why Charon (2005) argues that story telling is essential for treating pain and suffering: "illness calls forth the self… and the self is knowable only through stories." Charon coined the term "narrative medicine", and she regularly trains healthcare professionals in the use of narrative methods, especially in the treatment of those burdened with chronic pain and suffering.

While in medicine most insight-bearing narratives are spoken, written stories are helpful for difficult diagnoses. The physician or therapist may not be able to understand how to best interpret the essential facts and related events without writing up his or her own notes and then studying them. Likewise, the person suffering may be asked to write; the sufferer may gain important new self-insights and provide new information for diagnosis. Charon (2005) believes these stories are central to effective pain treatment, underscoring the importance of close partnerships between patient and physician, of authenticity in both roles, and of recognition of the mutuality of suffering. Clinicians suffer as they empathize with their patients' suffering, and they, too, need to learn how to accept suffering, not deny it.

Still, healthcare professionals working with patients quickly learn that of the close relationship between suffering and quality of life (Niv 2005). Narratives add details about small adjustments or even radical shifts in one's quality of life. They also suggest how one's daily life, relationships, and social context increase or decrease the experience of suffering. Niv (2005) asserts that, because a huge (e.g., 20 %) share of the United States population suffers from chronic pain and suffering, most healthcare professionals take for granted an understanding of the quality of life changes among those treated for chronic pain or suffering.

R. E. Anderson, *Human Suffering and Quality of Life*, SpringerBriefs in Well-Being and Quality of Life Research, DOI: 10.1007/978-94-007-7669-2_2,

2.1.1 Authenticity of Narratives of Pain and Suffering

Unfortunately, narratives cannot always be taken at face value. They may be falsified or exaggerated for various purposes. Morphine-based pain medications and other pain reducing drugs are illegal without a valid prescription in many jurisdictions, and they have a high street value. Sometimes scammers will use stories to make fraudulent insurance claims and get prescriptions from doctors.

Other fraudulent narratives are fabricated memoirs. A few years ago, several best-selling memoir authors were exposed as having embellished stories of great pain and suffering that were presented as 'true.' Essentially, the wide audience for such stories created a genre called 'misery lit.' Book publishers and television producers have taken advantage of the morbid motivation of many consumers to identify with others who suffer from pain and other calamities, and some have overlooked fraudulent narratives.

Illouz (2003) conducted an extensive analysis of Oprah Winfrey's talk shows and concluded that Oprah had created a "multilayered textual structure that initiates, stages and performs narratives of suffering and self-improvement, resonating with a wide audience." Using sophisticated discourse analysis, Illouz shows how Oprah used stories such as celebrity tragedies to not only entertain her audiences but also help them make sense of suffering. The stories were meant to give viewers an illusion of being 'on track' toward a more authentic identity. One tactic in this message, Illouz asserts, is the story of victim culture, which legitimizes large payouts by insurance companies and other institutions. This victim culture anesthetizes us to character corrosion (Senett 2000) and makes fabrications seem less than harmful. They may even seem like a form of 'deeper truth.'

Thus, while there are both personal and institutional pressures for contemporary stories to use suffering in inauthentic ways, the actual prevalence of authentic pain and suffering remains high, even growing. Surveys of pain and suffering reveal tremendous suffering worldwide, including in the wealthier nations.

2.2 Narratives of Suffering on the Internet

Because of its enormous storage capacity and active use by over 85 % of people in most age groups, the Internet, commonly known as the Web, has become a microcosm of human society. Its content represents most of the dominant cultures in developed societies. A large number of websites actively encourage sharing and reading stories.

With the help of several research assistants, my assistants and I searched the Web for any sites with stories about suffering. The selection of websites was limited to those that solicit stories related to pain or suffering. Thus, we excluded those sites with only a forum, where short comments and message exchanges are encouraged. We also eliminated sites catering to creative short story writing to

Table 2.1 Websites sampled for narratives of suffering

Websites with suffering narratives	Genres	Google links in 1,000s*
caringbridge.org	Bridging the suffering with friends	1,780
csn.cancer.org	Information and support groups	1,720
experienceproject.com	Life stories	8,860
goodtherapy.org	Faith, ethical therapy	70
helium.com	Writing feedback	52,100
lifestory.org	Life stories, faith	65,200
mayoclinic.org	Patient stories	3,729
mdjunction.com	Health information	1,020
somethingtoshare.com	Life stories, inspiration	2,220
save.org	Suicide-related stories, depression stories	2,120,000
suffering.net	Stories of pain, suffering, and faith	109,000
thereasons.ca	Suicide-related stories	245,000
voices.yahoo.com	Stories of suffering	112,000
whitewreath.com	Suicide-related stories	820
whypain.org	Stories of pain, affliction, and faith	118,000
Total		2,729,519

[a]*'Google links,'* in the rightmost column, is a measure of popularity or embeddedness within the Web. It is an estimate of the number of links to each of the above websites. It should be noted that this measure is limited, because some of the above websites have resources other than stories to which links may point

minimize the selection of inauthentic stories. Facebook has a "sufferings community" with 40,191 likes, and searching news sites yields many stories of suffering, but neither was included in this sample of sites. The sites we studied are listed in Table 2.1.

We randomly sampled stories, analyzing each in terms of its overall narrative and individual themes. Forty-five stories were analyzed, and the results are described and discussed within each of the eight frames of suffering.

Perhaps what stood out most was the raw pain and suffering described by the authors. In most website contexts, stories such as these received comments from others, and the original authors usually replied to acknowledge the comment. In a few instances, the comments told new stories that deserved mention.

Here is how one story, from the website *experience.com*, describes the agony of living with chronic pain. The writer, a young man in his 20 s, used the pseudonym 'reallybored.' While he did not give the source of his constant, piercing pain, he had been suffering for at least several months. He told his story in lyrical form, almost like rap music, which he explained: "I'm going out of my mind, the only thing I've got left is my ability to rhyme." Elsewhere, he wrote:

> I start to pour all my feelings out onto this page instead of punching my door; I'm on the verge of breaking. I can't keep faking being all right when I don't sleep at night.

He wrote that he had once been a musician, but could no longer play his guitar, which "felt like a knife." Not surprisingly, his story received several comments, which he answered politely.

2.2.1 Suffering as Punishment

The first frame, *suffering as punishment*, is not a generally popular, modern way to express the experience of suffering. Only one story we found explicitly referred to suffering served as a kind of punishment. Even the victim of the suffering had mixed feelings about the validity of punishment as the essential meaning of his or her suffering. The story's website was *save.org*, and the pseudonym 'anonymous.' Intense suffering began for this person 26 years before he or she wrote the story. Over these years, sleepless nights and turmoil continued despite a variety of treatments that included electroconvulsive therapy as well as numerous medications. The author did not share a specific diagnosis, but wrote of anxiety attacks and seven suicide attempts. At a younger age, a spouse, job, and apartment were lost. Religion seemed to have been both a source of comfort and internal struggles. At one point in the story, the victim said:

> I believed for a while that God was punishing me for something done in my past. Others said I was selfish and said things like "Just pull yourself up and out of it" and "Stop feeling sorry for yourself," which made me put myself down even more.

The story also gave evidence that this person used religious beliefs as a way to minimize suffering. The author wrote that "comfort was found in going to a safe place of worship," even as friends admonished him or her for what must have looked like the author's wallowing in pain. The storyteller's concluding advice to others was: "Know that you are worth living and that God loves us all, no matter what you have heard or what society says."

Even those who did not mention suffering as punishment may have attributed their pain and suffering to their past actions. The narratives we read were filled with family disruption, depression, social isolation, and so forth, all apparently direct consequences of the suffering. While these can be considered negative outcomes, they can also be conceptualized as components *of* suffering, but, as our objective was not to focus on causal consequences, that distinction was not explored in this analysis.

2.2.2 Suffering as Reward

The second frame, *suffering as reward*, consists of thinking about benefits gained from suffering. These might be small changes in one's thoughts and values, like increased gratitude, or they may be large shifts—some may believe withstanding pain and suffering is a desired religious experience.

Here is a story by Kattarrin on *experience.com*, written as an attempt to comfort 'reallybored'. According to Kattarrin's short story, she suffered from excruciating chronic pain and a seizure disorder that kept her from going out. If not for her daughter, she would not get out of bed. Still, she described suffering as a kind of reward:

> Someday I might find someone who will benefit from my story and my pain, even if it is only to know that they too are not alone. I am here for you and I will pray for you. May the Higher Powers hear you and bless.

Not only did Kattarrin hint that personal benefit might result from her extreme pain and suffering, but she explicitly stated that her agony might benefit others. 'Reallybored' could now know that at least one other person cared. Her words of comfort were both complex and very personal. She believed that intimate sharing or co-suffering gave her struggle meaning.

What Kattarrin was talking about is what is often called the 'all in the same boat' effect, where people sharing common grief or enduring a painful disaster together feel a common bond. This may lead to long term cohesion among those suffering together, which might partially compensate for the agony felt by each individual.

Another story, this time by Shannon on *thereasons.ca*, illustrates how suffering from calamitous events can be transformed into a reward or benefit. Shannon and her new boyfriend Michael seemed very compatible and happy together. Because of that, she did not worry much when he told her about bouts of depression and drug use. They both had suicidal thoughts earlier in their lives, but that was all in the past.

After a wonderful Saturday evening date, however, he talked about fighting at home, but he promised her not to push his parents. The next morning, Michael's sister called Shannon to say that he had committed suicide with a gun after an argument with his mother over what clothes to wear to church. Shannon was so devastated she wouldn't eat for a week and had to be hospitalized.

When her boyfriend's mother told Shannon that the last thing he held in his hand was a picture of Shannon, it was just enough to help her begin to restore her self-esteem. She explained it this way:

> I know that I must pull through for all the ones that love me, including Michael. He will always have a place in my heart and I know no one can ever replace the way I felt about him, but it was his time to go and God is using this story to help other people and impact their lives.

Shannon began to recover from her deep sorrow and suffering when she knew Michael's family had become aware of their dysfunctional role in his suffering. She grew closer to her friends and family, who became quite supportive, despite her initial phase of extreme withdrawal. People like Shannon, who are, for the first time, able to survive the death of a loved one or the agony of a painful illness, may ironically experience gratitude and a boost in self-esteem from having survived the tremendous emotional upheaval of a tragedy. They have come through something they previously could not envision tolerating.

2.2.3 Suffering as Craving

The frame for *suffering as craving* (or egocentric desire) is a popular attitude toward suffering in Eastern cultures. Equivalent notions of 'addiction as evil' and 'sin as intoxicating pleasure' are commonly held beliefs in most religious traditions. Buddhism teaches "Pain is inevitable; suffering is optional" and "the origin of suffering is craving," and, in Western religions, suffering is sometimes said to result from indulging in 'self-centered pursuits of the flesh.' The implication is that exerting self-control over these immoral urges will free one from suffering.

Rick Derringer is an American rock star, having recorded 22 albums over past 35 years and performed with famous bands from the USA and Britain. He added a story to the website *lifestory.org*. In addition to listing his celebrity experiences, he admitted his addiction to alcohol and drugs. As he began to feel like his life was falling apart, he wrote, he remembered his Catholic upbringing and returned to the practice of prayer. His story states:

> The Lord allowed me to survive drugs, alcohol and sins of the flesh so that I can stand here today as an example… The Lord can bless you with the strength to beat any addiction.

The implicit message here and in numerous other Christian testimonies is that human suffering is a direct consequence of cravings of the 'flesh,' a metaphor for self-absorption or attachment to anything that brings self-centered pleasure. It follows, in most religious teaching, that the way to escape from suffering is to stop one's craving for egocentric or 'sinful pleasures.' The weakness of this point of view is that most people need more motivation than the possibility of vague suffering to give up things they and their friends think they enjoy. A more effective viewpoint is that alternative, other-centered pleasures offer deeper, long lasting forms of pleasure.

2.2.4 Suffering from Sacrifice or Altruistic Actions

This is sometimes called compassion fatigue or empathy fatigue. Stories about the suffering that results from compassionate action or empathic altruism tend to be scarce. Perhaps altruistic action is relatively rare. Another possibility is that people engage in prosocial or helping behavior, but do not think of it as altruistic or compassionate. More likely, people engage in altruistic actions, but only when it does not require a large sacrifice on their part. So, whether an act requires great self-sacrifice—enough to produce suffering—may depend on how much one values doing things for others. Extreme altruism will likely result in some degree of suffering on the part of the giver, but this is not the case for a great deal of caregiving.

The following story was found on *suffering.net*. A young boy's older brother was severely injured in a car crash. He needed a blood transfusion, and his younger brother had the only blood type that matched. Their father asked the boy if he would be willing to give his blood so that his older brother could live.

Without hesitation, he agreed. Later, after a vial of his blood had been drawn by a nurse, the little boy turned to his father and asked:

> Daddy, how long do I now have before I die?

One vial of blood would not kill him, but the boy did not know that. He believed his gift to his brother would mean giving up his own life, and yet he had not wavered. As stated in the New Testament (John 15:3), "Greater love has no one than this, that he lay down his life for his friends." Rarely are we asked to sacrifice to this point of major suffering. Yet to end severe suffering around the globe may require most of us to sacrifice or suffer far more than we do now. This frame on suffering deserves thoughtful reflection by all.

Perhaps the most important conclusion is that genuine empathy and compassion do not demand that we suffer to the same degree as those suffering. They can be helped and their suffering relieved without actually experiencing the depth of their 'unbearable' suffering.

2.3 Suffering as Natural Destiny

Depending upon the time and place, much of the world's suffering is caused by events of nature or random forces that we cannot control. Such destiny or fate is certainly true for natural disasters such as earthquakes. Other calamities such as floods, epidemics, famines and weather storms may be the consequence of both natural forces and human actions. We have few words in the English language that help us to think about the forces of nature that pain and suffering produce, but 'destiny' and 'fate' are two. Pain and misery are inevitable; taking undue blame for them is not and should be avoided whenever possible. Likewise, we should not blame fate for disasters produced by human behaviors. Therefore, it is important to learn more about natural disasters so that we can distinguish 'acts of God' from events resulting from technology or other human creations. By allocating blame correctly, we can engage in actions that will alleviate or perhaps avoid more human misery than already exists.

Brandi, on *stories.yahoo.com*, wrote a vivid description of her first miscarriage. She had had one successful child birth, but this was totally different. The ordeal of severe pain and unsuccessful trips to the hospital continued for several weeks. When she finally miscarried at nearly seven weeks of pregnancy, the pain was excruciating and the grief overpowering. Three more miscarriages followed during the next year. With each new miscarriage, her pain and suffering continued. Eventually, Brandi came to peace with the repetitive ordeal. This is how Brandi explained it:

> I now know that most miscarriages are due to something being wrong with the baby and miscarriage is nature's way of letting only the strong survive.

Coming to believe that miscarriages are "nature's way" of eliminating embryos with genetic problems was eye-opening for Brandi, making it much easier to accept her pain and cope with her grief. Something as simple as learning the scientific facts behind suffering, or seeing more clearly the role of destiny, can greatly reduce needless suffering.

2.4 Suffering as Pain Management

The most frequent narratives on the management of pain and suffering were off-handed comments about the inadequacy of the authors' pain medications. Quite a number of the online writers referred to their pain meds as a joke; they helped, sure, but the help was tiny compared to the pain remaining.

As expected, those suffering from cancer were especially unhappy with their medications. They talked about their chemo meds as making them sick, but felt they had no choice but to take them. Several even referred to their medications as dangerous, perhaps because the patients' were sometimes left with severe pain but also because taking the medications often produced new, unpleasant side effects.

A third common theme was the stigma that came with taking potentially habit-forming drugs for pain. A woman with the pseudonym 'actvforlife' at *experienceproject.com* wrote of severe and constant neck and back pain stemming from a whiplash injury in an auto-truck crash. After 6 years, she on medication, and while it helped, it made her feel alienated:

> I am on pain meds, which are a stigma for me from my family.

Medications are not the only tools for pain management. Hypnosis and alternative medicine techniques including dietary adjustments may help manage pain. One strategy some of the story writers shared was improved self-compassion (although this is my interpretation). Neff (2011), a leader in the self-compassion movement, defines self-compassion as embracing one's pain, feeling unconditional self-kindness, and avoiding destructive patterns of fear, negativity, and isolation. Research on the effectiveness of self-compassion training confirms its effectiveness in the management of pain and suffering (Neff 2011). Although pain management is generally associated with prescription drugs and medical procedures, there are spiritual practices like meditation, and self-compassion (as opposed to self-pity) exercises that should be included among the tools for the management of pain.

2.4.1 Relief of Suffering as Human Purpose

Mayerfeld (1999) persuasively argues that the reduction of suffering is morally more important than the promotion of happiness and that "most of us greatly underestimate the force of the duty to prevent suffering." There are numerous ways to work toward the relief of suffering, ranging from caregiving for a family member (perhaps a sick child or an aging parent) to becoming a full time disaster relief worker. Such dedication benefits not only the victims of the suffering, but also the reliever of that suffering—more commonly called the caregiver. Caregiving benefits the caregiver primarily by providing a highly compelling purpose or meaning for living. This meaning often is reinforced by the gratitude of those whose suffering is abated.

The experiences of suffering persons and caregivers have been investigated extensively among those who volunteer as within-family caregivers. Some of the findings from this research, both quantitative and qualitative, are mentioned here because they illustrate the benefits and costs that come to those who work toward the reduction of suffering.

The American Cancer Society estimates that three out of four families have at least one member who is a cancer survivor. Their National Quality of Life Survey for Caregivers began in 2002, when 1,635 cancer caregivers were surveyed. The study followed up in 2005 and 2008. Many, but not all, caregivers reported their cancer caregiving experience as having given them new perspectives and added meaning in their lives.

Based on this study, Kim et al. (2007) concluded that multiple personal pathways helped caregivers "find meaning in their caregiving role." The study showed that caregivers who found meaning in the experience developed greater satisfaction with life and experienced fewer episodes of depression. These results were confirmed in a European study by Minaya et al. (2012).

In another study (not limited to cancer care), the Washington State Personal Family Caregiver Survey discovered that many family caregivers find deep satisfaction and meaning in their role, but it often comes at substantial cost to the caregiver's own physical and mental well-being. Researchers Montgomery et al. (2007) concluded that, while some caregivers derived greater meaning from their lives as caregivers, they also struggled with identity change, as did all long-term family caregivers.

Noonan et al. (1996) interviewed 48 informal caregivers to the elderly, and the narratives they reported support others' claims about the importance of meanings and the quality of life in understanding suffering. They found that predominant caregiving mentioned included gratification and satisfaction, family responsibility and reciprocity, and friendship and company.

Here are some illustrative quotes:

> My aim is to make the quality of her life, what she has left, as nice as it can be.
>
> Mother is happier than she's ever been in her life, which makes me very happy.
>
> I'm doing the best I can and am glad to do it… It is a labor of love.

Finally, while the authors of our collection of web stories were not caregivers themselves, many expressed a hope that others learn from their experiences of agony and misery. Sabrina, who left a story on *whitewreath.com*, is a good example. After struggling for years with suicide attempts, she recovered and now provides help for others with online comments and suggestions. One such suggestion was:

> One day you will realize that you don't have to hide, you will realize you are much stronger than you ever gave yourself credit for being.

Those who have experienced the agony of suffering themselves and put it behind them (or, at least, have come to manage it better) tend to feel empathy for those suffering in similar ways. Empathy does not make one a therapist, of course, but these people may be able to inspire and support others in proactive approaches to their challenges.

2.4.2 Relief of Social Suffering as Progress in Quality of Life

As we found no web stories related to this frame of suffering in the 15 websites of our main source of narrative data, special searches were made for stories about humanitarian aid and relief workers and the meaning of their work. The principal conclusion of this extensive search was that humanitarian aid and relief workers almost never write about their own motivations for reducing suffering. Perhaps they have been neglected as a professional group because the organizations for which they work put their time and attention toward the huge and horrendous plight of their clientele, the suffering victims of various calamities. This may leave relief workers less likely to write about their own comparatively 'unremarkable' experiences.

It goes without saying, though, that reducing or eliminating the suffering of victims of tragic disasters improves the victims' quality of life, to the extent that they make contact with humanitarian aid of some kind. How quickly and extensively their quality of life improves depends largely upon the funding support and organizational effectiveness of humanitarian aid and relief programs. A large part of program effectiveness is the human capital represented by humanitarian relief professionals, so it is surprising that so little is known about them.

One major exception is a book (Bergman 2009) that contains a large collection of stories by aid and relief workers. These stories give us glimpses into workers' motivations and the meanings of their work supports their lives. In the introduction, Bergman wrote:

> Humanitarian workers, in general, have a different notion of home and security. They often complain, jokingly, of the pressures of a "normal" life and admit to enjoying, or needing, the adrenaline rush of the front lines.

One of his stories is exemplary:

> I continue working. I do this type of work because I believe in what we are trying to achieve and experience a tremendous sense of satisfaction when I help people. I see severely malnourished children on one trip, and when I return a month later, they are running around and smiling. (p. 132)

Another story Bergman (2009) included is by a Vietnamese woman who worked as an aid worker. In the era after the Vietnam War, while flying over the country, she observed:

> No longer suffering from war, they're suffering from a different kind of struggle—fleeing from poverty and natural disasters. (p. 39)

Her comment points out how the social suffering of a peaceful era may be as devastating as civil war. If not won, the war on poverty and environmental preservation can lead to as much destruction of human life as a major, modern war from which an estimated four million people died.

Another perspective on the meaning of aid work to an aid-worker is summed up in a quote from a long eBook written by Peter (2007) about his life as an aid worker:

> This could have been my family, my life. But destiny has put them there and me here. Sheer luck determined those who suffer and those who never realize how lucky they are. Sheer destiny determined those who need help and those that can help. I can help. And that is why I am an aid worker.

Reduction of suffering is what makes life most meaningful for someone devoted to humanitarian action. But reduction of suffering is not enough. All people must also work to reduce social suffering, finding and eradicating the root causes of poverty, violence, and other social forces that fuel greater and greater suffering.

2.5 Conclusions

The culture of the Web provides for the emergence of an online genre of short storytelling that magnifies drama. Gripping stories may capture the attention of readers who have had similar experiences or who can empathize with the author's suffering. The typical story exudes deeply felt, negative emotions associated with traumas such as a bout of depression, a suicide attempt, a late-stage abortion, major illness, or pain due to accident-induced muscle damage.

One of the most common sources of suffering in Web stories is depression, usually clinically diagnosed. The suffering we read about ranged in length from a few weeks to 75 years of depressive episodes. The most extreme depression in the Web stories was that of someone who lived with bipolar disorder over a period of years, including physical and mental pain and attempts at suicide. The author's words paint a portrait of a life that feels like war-time torture.

Another qualitatively different type of story is that of a mother grieving for the loss of her son to suicide. Though he had been diagnosed with schizophrenia and depression, she still seemed to shoulder some of the blame for his death. Though a relatively mild instance of suffering, she admitted that the episode degraded her quality of life because now she is "basically scared of everything."

While some suffering arises from a shared calamity, the Web stories of suffering were typically about personal trauma. Parallel suffering, such as commonly experienced after a major earthquake, tends to provide a natural social support system. Alexander's (2012) analysis of social traumas reveals how unbelievably tragic and horror-filled many stories of cultural trauma end.

In the Web stories, the victims' family and other support systems sometimes failed to provide adequate care and comfort. Chapman and Volinn (2005), in their study of chronic back pain, found that victims often experienced serious problems with family disruption. Some of this arose around the sufferer's inability to work and provide their former level of income.

Other problems arose because family members experienced ambiguous loss (Boss 1999; Mulvany 2010). Such loss arises when it is difficult to predict if (and when) the victim might return to their former participation in family relationships. Role reversals, changed roles, and withdrawal from family activities can all threaten the cohesion and viability of family and individual relationships.

References

Alexander, J. C. (2012). *Trauma: A social theory*. Cambridge: Polity Press.

Bergman, C. (2009). *Another day in paradise: International humanitarian workers tell their stories*. Eugene: Wipf & Stock.

Boss, P. (1999). *Ambiguous loss*. Cambridge: Harvard University Press.

Chapman, C. R., & Volinn, E. (2005). Narrative as a window on chronic disabling back pain. In D. B. Carr, J. Loeser, & D. B. Morris (Eds.), *Narrative, pain, and suffering* (pp. 73–86). Seattle: International Association for the Study of Pain Press.

Charon, R. (2005). A narrative medicine for pain. In J. Carr, D. Loeser, & D. B. Morris (Eds.), *Narrative, pain and suffering* (pp. 45–53). Seattle: International Association for the Study of Pain Press.

Illouz, E. (2003). *Oprah winfrey and the glamour of misery*. New York: Columbia University Press.

Kim, Y., Schulz, R., & Carver, C. S. (2007). Benefit finding in the cancer caregiving experience. *Psychosomatic Medicine, 69*, 283–291.

Minaya, P., Baumstarck, K., Berbis, J., et al. (2012). The CareGiver oncology quality of life questionnaire (CarGOQoL): Development and validation of an instrument to measure the quality of life of the caregivers of patients with cancer. *European Journal of Cancer, 48*(6), 904–911.

Montgomery, R. J. V., Rowe, J. M., & Kosloski, K. (2007). Family Caregiving. In J. A. Blackburn & C. N. Dulmus (Eds.), *Handbook of gerontology: Evidence-based approaches to theory, practice, and policy*. New York: Wiley.

Mulvany, A. P. (2010). *Flood of memories: Narratives of flood and loss in Tamil South India*. Publicly accessible Penn Dissertations. Paper 383. http://repository.upenn.edu/edissertations/383. Accessed 18 May 2013.

Neff, K. (2011). *Self-compassion: Stop beating yourself up and leave insecurity behind*. New York: HarperCollins.

Niv, D. (2005). The chronic pain narrative and quality of life. In D. B. Carr, J. D. Loeser, & D. B. Morris (Eds.), *Narrative, pain, and suffering* (pp. 53–72). Seattle: International Association for the Study of Pain Press.

Noonan, A. E., Tennstedt, S. L., & Rebelsky, F. G. (1996). Making the best of it: Themes of meaning among informal caregivers to elderly. *Journal of Aging Studies, 10*(4), 313–327.

Peter, (2007). Why I am a humanitarian aid worker. In eBook, *The road to the horizon*. http://www.theroadtothehorizon.org/2009/03/why-i-am-humanitarian-aid-worker.html. Accessed 13 Feb 2013. Note: the author identifies only as 'Peter'.

Sennett, R. (2000). *The corrosion of character: The personal consequences of work in the new capitalism*. New York: W. W. Norton.

Chapter 3
Statistical Portrait of Suffering in America

The previously discussed taxonomy of suffering gives structure to the concept by dividing it into three major types of suffering: physical suffering or pain, mental suffering, and social suffering. Statistics from a large, national health survey reported here provide a portrait of suffering in America. Physical suffering, typified by chronic pain, usually depends upon neurological paths between a sensory organ and the brain, as a communication system. However, recent neuroscience research has discovered a number of ways that pain is instigated without following the simple neurological pathways (Borsook 2012).

Mental suffering does not necessarily have an origin in painful sensory events, and is more elusive. Depression and anxiety, perhaps the most common varieties of mental suffering, when combined with other mental maladies like grief and existential suffering, together form a major type of suffering labeled here as mental suffering. The third type of suffering, known as 'social suffering', is a relatively new label for suffering that is produced primarily by social conditions that damage a collectivity's sense of self-worth and heightens powerlessness produced from socially shared traumas. One consequence of social suffering often is the loss of caring for self and others as valued human beings.

Narratives of those who suffer uncover experiences that range from minor hurts to agonizing, life-threatening events. Such experiences also differ in their persistence over time and the meanings superimposed on them. Not surprisingly, stories of suffering often raise the question of how often different types of suffering occur, and how much severe suffering exists in the world.

These are not easy questions to answer because the measurement of suffering in healthcare and public opinion surveys remains relatively primitive. It does not help that the measurement of suffering has not been a priority for either the public or private sectors. Now that pain management has come to play such a central role in healthcare, the measurement of pain has received more attention (Cassell 2004).

R. E. Anderson, *Human Suffering and Quality of Life*, SpringerBriefs in Well-Being and Quality of Life Research, DOI: 10.1007/978-94-007-7669-2_3, © The Author(s) 2014

3.1 Data for Measuring Suffering in the United States

The data used in this profile of suffering in the United States was collected as part of the National Health Interview Survey (NHIS), the longest-running health survey in the world. Approximately each year beginning with 1957, the NHIS surveyed from 50, 000 to 100,000 randomly selected Americans. Using a complex sampling design, the data collected each year represent a scientifically selected sample of the non-institutionalized adult public. These data are used to monitor the health of the US population, track health progress, and evaluate the quality of healthcare in the United States. NHIS is designed by the CDC's National Center for Health Statistics (NCHS)–the government agency tasked to monitor the population's health status, and administered by the US Census Bureau.

The NHIS data used in this study were obtained from the Integrated H0065alth Interview Series (IHIS) database (available at http://www.ihis.us). The data are managed by the Integrated Public Use Microdata Series (IPUMS) at the University of Minnesota Population Center. The project was funded by a grant from the National Institute of Child Health and Human Development (NICHD). More details about both the NHIS and the IHIS projects are given in Johnson et al. (2008).

All of the data used in this analysis were collected in the year 2010, in which approximately 66,000 adults across the United States were interviewed in person. The response rate at over 90 % was very high for such a large study. Not all of the respondents were asked all the questions, because blocks of questions were administered to carefully designed subsets of people in order to avoid any one person to have to answer hundreds and hundreds of questions.

Many of the suffering-related items used in this analysis were asked of only 6,115 respondents; however, these were randomly selected within randomly selected geographic clusters, in accord with the complex sampling design. This complex sampling design was taken into account when estimating the accuracy of the estimates in terms of the confidence intervals calculated and the differences reported. The population of non-institutionalized adults in 2010 was approximately 229,502,464. This was the percentage base used in our analysis, less any missing responses due to refusals or lack of information.

3.2 Indicators of Specific Types of Suffering

Some trivialize suffering because they are unaware of how pervasive suffering has become even in wealthy societies. The approach taken here is this by breaking different types of suffering into measureable components, it becomes possible to quantify suffering overall and more precisely identify its depth and scope. We begin with the type of suffering that has received the most investigation: pain, sometimes called physical suffering.

3.2.1 Physical Suffering

Epidemiological studies often focus on 'chronic pain,' which is a medical term, but one that is defined in many different ways. Health care professionals and researchers generally agree that pain is chronic when a person feels it over at least a 3 or 6 month period, however, no consensus has yet emerged on the degree of severity or how often it must occur for pain to be chronic. A 3-month minimum time frames is more popular, perhaps because, as pointed out by the American Academy of Pain Medicine (2013), it takes about 3-months for tissue damage to naturally heal.

Hardt et al. (2008) analyzed the 2000–2003 NHANES (National Health and Nutrition Examination Survey) data and reported a chronic pain level of 10 % for chronic pain among adults in the USA. Johannes, Le, Zhou, The Institute of Medicine (2011) gave an estimate of 100 million adults in the USA or a 45 % prevalence rate of chronic pain. Johnston and Dworkinin's (2010) recent Internet survey of chronic pain reported a 31 % prevalence rate for chronic pain in the USA. It is important to note that the study had a relatively poor response rate of 45 % and the data collection mode was that of an internet survey.

The lack of consistency in definitions of chronic pain implicit in the questions asked of the respondents account for the wide range in prevalence rates for chronic pain. Furthermore, because the audience of these surveys is generally a medical community with a special interest in pain in a particular body area, the questions typically asked about an individual source of pain rather than the general qualities of the pain itself.

This problem undoubtedly contributes to the wide variability in international comparisons of chronic pain prevalence, which range from 10 to 55 %. Harstall and Ospina (2003) evaluated 13 major studies of chronic pain in Europe, Canada, Australia, and Israel, of which about one-third defined chronic pain as lasting at least 6 months and the remainder, 3 months. Across these 13 large studies, the prevalence of chronic pain ranged from 10 to 50 % in adult populations. Across all 13 studies, the weighted average was about 31 %, suggesting that almost a third of the populations across developed countries reported to be suffering from 'chronic pain.'

Tsang et al. (2008) reviewed pain surveys in 18 countries, about half of which were highly developed and the remainder underdeveloped. Both groupings of countries had an average chronic pain prevalence of about 30 %. When chronic pain was combined with the prevalence of depression and anxiety, the prevalence of the combined three types of suffering was not significantly different for developing versus developed countries.

In the IHIS study reported below, chronic pain was operationally defined using the IHIS data in 2010 as the occurrence of pain that people reported as having 'most days' or 'every day' over the previous 3 months. See Table 3.1 for the exact wording of the question asked in IHIS regarding how often one's pain occurred. If someone said, his or her pain re-occurred 'every day' or 'most days,'

Table 3.1 Suffering indicator descriptions for IHIS 2010 analysis

Social indicator name	Operationalization procedures and definitions	IHIS variables names and codes used
Physical suffering (also called chronic pain)	Chronic pain was considered present when the respondent (R) reported 'every day' or 'most days' in response to this question: "In the past 3 months, how often did you have pain? would you say never, some days, most days, or every day?" also called PHYSICAL SUFFERING	Painfreq3mo = 3 or 4
Extreme pain	Extreme pain was considered present when the R reported (1) 'chronic pain' (see above) (2) 'frequent' pain (3) that the pain was 'sometimes unbearable and excruciating,' (4) that the pain was 'constantly present,' (5) disagreement with the statement "Medication can take my pain away," (6) disagreement with the statement "When I get my mind on other things, I am not aware of the pain." All six conditions were required for classifying the pain as 'extreme'	Painfreq3mo = 3,4 and Painfreq = 2 & painintense = 2 & painconstant = 2 &painrxcure = 1 and Paindistrac = 1
Anxiety	R reported that he/she felt 'worried, nervous or anxious every day'	Worfreq = 1
Extreme anxiety	R was classified as having daily anxiety (see anxiety) and R agreed with the statement "Sometimes the feelings can be so intense that my chest hurts and I have trouble breathing"	Worfreq = 1 & worrx = 2 & anxintense = 2
Depression	R answered 'daily' or 'weekly' to the question: "How often do you feel depressed? Would you say daily, weekly, monthly, a few times a year, or never?"	Depfreq = 1,2
Extreme depression	R was classified as having depression (see DEPRESSION above) *and* R agreed with the statement: "Sometimes the feelings can be so intense that I cannot get out of bed"	Depfreq = 1,2 & depintense = 2
Grief	R reported that his or her "feelings of depression were caused by death of a loved one"	Depymourn = 2
Existential suffering	Respondent had at least one of these three conditions all of the time in the past three 3 months: (1) feeling hopeless (2) feeling worthless or (3) feeling sad	Ahopeless = 4 or aworthless = 4 or asad = 4
Mental suffering	For any R, meeting the criteria for at least one of the following suffering types: anxiety, depression, grief, or existential suffering.	
Social suffering	Any instance of one of these four factors: poverty, disability, blindness, or deafness (see Table 3.2 for definitions). Value set to one if below poverty line, mobility disability, blind, or deaf, else 0 (zero)	Pooryn = 2 or disability = 1 or Blindess = 1 or deafness = 1
Any suffering	Any instance of physical, mental or social suffering.	
Physical or mental	Any instance of either physical or mental suffering.	

Abbreviations: *IHIS* International Health Interview SStudy; *R* Respondent

Table 3.2 Suffering estimates for indicators of adult USA Population in 2010

Social indicator name	Prevalence in percent of adults (%)	Prevalence in population count	Non-missing cases in sample	95 % conf. interval, lower bound (%)	95 % conf. interval, upper bound (%)
Physical suffering (also called chronic pain)	19	39,412,455	6,123	17.7	20.3
Extreme pain	3.3	6,895,615	6,115	2.8	3.9
Anxiety	8.4	17,567,148	26,977	7.5	9.4
Extreme anxiety	4.5	9,519,683	6,141	3.9	5.3
Depression	9.2	19,146,689	6,115	8.4	10.1
Extreme depression	5.4	11,301,555	6,115	4.8	6.1
Grief	2.8	5,872,474	6,117	2.3	3.3
Existential suffering	4.8	11,043,818	7,195	4.2	5.5
Mental suffering	14	29,824,789	24,787	13.0	15.2
Social suffering	17	34,531,074	22,943	16.2	18.7
Any suffering	35.7	69,033,544	6,115	32.1	36.8
Physical or mental	27.1	52,869,953	6,123	25.7	28.5

Source All the data reported in these tables were extracted from the International Health Interview Survey (IHIS) for 2010. The data were collected in the National Health Interview Survey (NHIS) and are publically available from Integrated Public Use Microdata Series (IPUMS) at the University of Minnesota Population Center

then she/he was categorized as having been in chronic pain. Using this criterion, 19 % (or about 40 million) of the United States adult population had chronic pain in 2010. In this report, we use the terms 'chronic pain' and 'physical suffering' interchangeably.

For this and all the other indicators of suffering prevalence reported here, Table 3.2 gives the prevalence statistics and their confidence intervals, as well as the number of cases each indicator was based upon.

3.2.2 *Extreme Pain*

Severe pain which some people describe as traumatic, dreadfully hurtful, unrelenting, or debilitating varies by intensity and persistence. In this study, such pain is labeled 'extreme pain.' For present purposes, extreme pain has been operationally defined as having all six of these attributes: (1) frequent pain in the previous 3 months, specifically having felt pain on "most days" or "every day;" (2) recent intense pain (3) constantly present pain over the past 3 months (4) pain feeling "sometimes unbearable or excruciating;" (5) pain that does not go away when taking pain medication, and (6) pain that does not go away when one's mind is on other things. Using this strict standard, 3 % of the US adult population, or nearly seven million people self-report such extreme pain (See Table 3.3).

Each individual in the survey was also asked to rate their pain over the last seven days on a scale from 0 to 10 or from the absence of any pain to the highest

Table 3.3 Percent of americans with each type of suffering

Types of suffering	Percent of adults (%)
Any suffering	36
Physical or mental	27
Physical suffering	19
Social suffering	17
Mental suffering	14
Any extreme suffering	13
Depression	9
Anxiety	8
Either extreme anxiety or extreme depression	8
Extreme depression	5
Existential suffering	5
Extreme anxiety	5
Extreme pain	3
Grief	3

Source See Table 3.2 for details of data source

possible pain. Those qualifying for chronic pain on average rated their pain 'in the last week' as 5.3 whereas those in extreme pain rated their last week's pain as 7.3 on average. Even though the time frame of the pain rating question was more recent, the ratings show a very large difference in pain severity across the two groups, thus contributing toward validation of the pain classifications.

3.2.3 *Mental Suffering*

Mental suffering, as previously defined, refers to emotional and cognitive trauma largely separate from physical suffering. In this study, mental suffering was defined as present when a person qualified as possessing one or more of the following four types of mental suffering: Anxiety, Depression, Grief, and Existential Suffering. 14 % of (about 31 million) adults admitted to having at least one of these types of mental suffering and hence were classified as "mental suffering."

3.2.4 *Anxiety*

Anxiety encompasses prolonged worry and restlessness and what is sometimes called "nervousness." In this study, anxiety was operationally defined as instances when respondents admitted to feeling "worried, nervous or anxious every day." Using this criterion, 8 % of adults in the USA or 17.6 million Americans had anxiety.

3.2.5 Extreme Anxiety

Extreme anxiety includes conditions sometimes diagnosed as a disease or a state of mental illness. It also includes compulsive obsessive disorders and major addictions such as alcohol, drug, food, gambling and shopping addictions. For present purposes, extreme anxiety has to meet both the criterion for anxiety in general and a second criterion of admitting to a statement that sometimes my anxiety is so intense that "my chest hurts and I have trouble breathing." In this study, 4.5 % of American adults were classified as having extreme anxiety, which is an estimated 9.5 million adults in the USA that met these criteria.

3.2.6 Depression

Depression was assumed to exist, for present purposes, if a respondent gave an answer of 'daily' or 'weekly' to the question: "How often do you feel depressed? Would you say daily, weekly, monthly, a few times a year, or never?" 9 % met this criterion and thus were classified as having depression, which means that about 19 million adults in the USA suffered from depression in 2010 by this criterion.

3.2.7 Extreme Depression

Extreme depression refers to a more intense and broader form of depression. Any respondent was considered extremely depressive if he or she met all three of these conditions: (1) gave an answer of 'daily' or 'weekly' to the question: "How often do you feel depressed? Would you say daily, weekly, monthly, a few times a year, or never?" (2) felt so much depression that he or she agreed that to sometimes not being able to "get out of bed." and (3) reported feeling depressive every day. About five and a half percent or 11 million adults met this criterion for extreme depression.

3.2.8 Grief

A survey respondent was considered to be in a state of mourning or grief if she/he reported that his or her depression was caused by the death of a loved one." Three percent or roughly 6 million American adults self-reported themselves to be in a state of grief over the loss of a loved one.

3.2.9 Existential Suffering

Existential suffering consists of the absence or loss of meaning that is exhibited by indications of hopelessness, negative self-worth, loss of meaning, and spiritual or moral confusion (Langle 2008; Williams 2004). In the study described here, a respondent was classified as having existential suffering if s/he had at least one of these three conditions persistently throughout the past three 3 months: (1) feeling hopeless (2) feeling worthless or (3) feeling sad. Nearly five percent of adults, which adds up to about 11 million, qualified by these criteria as being in a state of existential suffering.

3.2.10 Social Suffering

Social suffering encompasses any major suffering that occurs in a social context and necessarily affects other people in a major, negative way (Kleinman et al. 1997). Thus, social suffering includes major disabilities, poverty, and other victims of racial discrimination or hate-based social targeting (Bourdieu, 2000). This study operationalizes social suffering by including anyone that is below the poverty income line or anyone with severely restricted physical mobility. Table 3.4 describes the indicators used to define this concept; the indicators were POVERTY and IMMOBILITY, respectively. The contexts of poverty and severe disability are very social. That is, both conditions are highly visible to others, shape a person's identity, and influence the people in these contexts highly vulnerable to being not only socially restrained but also stigmatized in the larger society (Macdonald and Jensen-Campbell 2010; Wilkinson 2005).

In the adult population, 17 % or 34,474,074 were categorized as "social suffering." This study contained other indicators that might have been used to refine or extend the social suffering categorization; however, the other indicators had technical problems such as the wording of the questions or an unusually small number of cases remaining in the sample. For this and the other suffering indicators, the operationalization has been logical and statistically valid even though exploratory. A summary of the prevalences discussed above are given in Table 3.3 ordered from largest to smallest prevalence.

So far, the discussion has not considered the fact that all of these types of suffering overlap with each other. The "extreme" types for pain, depression, and anxiety are subsets of the more general type. But the other types overlap with substantively different types of suffering, e.g., physical with mental suffering.

This pie chart (Fig. 3.1) shows the amount of the overlapping among the three major categories: physical, mental and social suffering. (The amount of the overlap was calculated by cross tabulating each type of suffering with every other type.) First, 36 % of the adult population reported some type(s) of suffering. Four percent of the population had all three types of suffering concurrently, and 12 % had two

Table 3.4 Demographic and quality of life indicator descriptions for IHIS 2010 analysis

Social indicator name	Operationalization procedures and definitions	IHIS variable names and codes
Poverty	Individual's household income in 2009 below or above official U.S. govertment. poverty guideline as defined by the office of management and budget (OMB). level depends on # persons and # children in household	Pooryn = 2 else 0
Immobility	Immobility determined if in answer to "are you able to carry out everyday physical activities?" R answers 'a little' or 'not at all'	Physactable
Age	Age in years since birth for those 18 years and older. age 85 includes all 85 and older	Age
Age55+	Dichotomy of age: 18–54 coded 0, and 55 and over coded 1 (see age above)	Age
Sex	Gender: men coded; women coded 2	Sex
Income	Total annual personal income in 2009 divided into 11 increments with the 1st as 0 to $4,999 and the last as $75,000 or more	Earnings
Qualoflife	For quality of life, R's were asked "in general, would you say your quality of life is excellent, very good, good, fair, or poor?" (coded 1–5)	Qol
Social support	Degree of satisfaction with one's social activities and relationship on a scale from 1 (excellent) to 5 (poor)	Socsatisfy
Overall health	R's self-assessed physical health status from excellent, very good, good, fair, to poor (codes 1–5)	Healthphys

Abbreviations: *IHIS* International Health Interview Study; *R* Respondent

types of suffering concurrently. Social suffering overlapped more than did physical and mental suffering. This may be a consequence of social suffering having greater influence upon the likelihood of getting physical or mental suffering than vice versa.

3.2.11 Any Suffering

To produce an overall estimate of the number (and share) of the population with any suffering, an indicator was created that identified those who had one or more types of suffering. As shown in Table 3.2 36 % (69 million) had physical, mental, or social suffering, or some combination of the three types. About half of the adult population had one or more types of suffering. The prevalence of each of the three suffering types was about equal.

A principal finding in this analysis is that there is a strong tendency for major types of suffering to occur together. Two-thirds of those who shoulder the distress

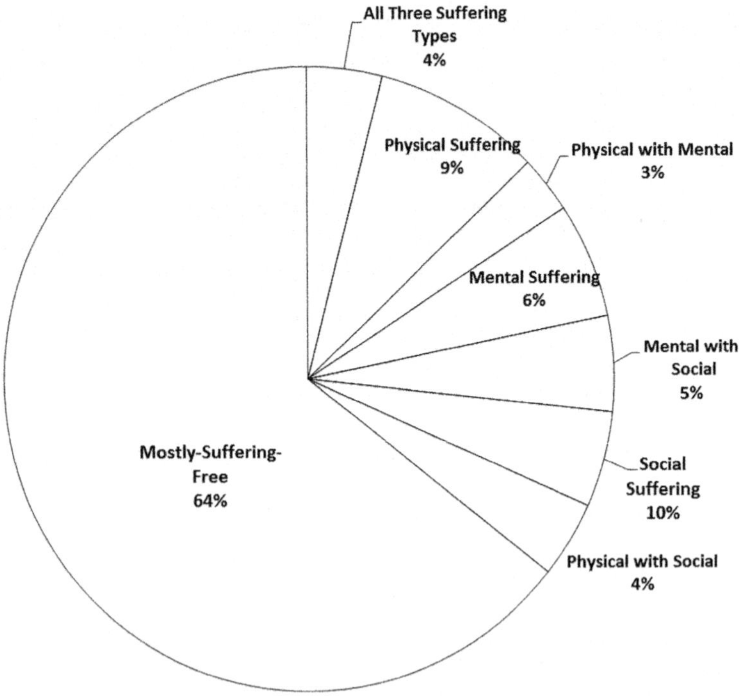

Fig. 3.1 Three types of suffering of USA adults in 2010 (IHIS)

of social suffering, also suffer from physical and/or mental suffering. In contrast, only 44 % of those who experience the trauma of physical or mental suffering also experience another type of suffering. The significance of this finding is that developing countries with a high degree of poverty and/or disability are likely to have a much higher incidence of all three types of suffering than a relatively wealthy society like the United States.

3.2.12 *Any Extreme Suffering*

Even if we limit our count of suffering persons to those who have '*extreme* suffering,' the numbers are startlingly high. 'Extreme suffering' is defined as anyone who described their experience over the previous three months as (1) pain that is "excruciating and unbearable," or (2) depression so bad that they sometimes "cannot get out of bed," or (3) their anxiety was so overwhelming that their "chest hurt" and they "had trouble breathing." I discovered that 13 % of Americans (25 million adults) struggle with that level of extreme suffering (see Table 3.3).

To some extent, such severe suffering is not preventable, but much of it is. For instance, if poverty were eliminated, nearly 5 million fewer American adults would have to live with extreme suffering. Or if we were to cut US adult obesity in half, 3.3 million fewer American adults would be in extreme suffering. This estimate presumes that obesity increases serious illness or injury, which in turn results in severe suffering. The interesting implication is that a lot of extreme suffering could be prevented by health and social policy. This knowledge offers a new perspective to address an old problem, extreme human suffering. The question is how much do we value the reduction of human suffering and what are we willing to sacrifice so that others may be given an opportunity to escape their suffering?

About half of those who shoulder the distress of any one type of suffering, also suffer from one or more other types. In some cases, they may be causally related. Their inter-relatedness suggests that they all prove burdensome for human beings and in extreme instances, they become incapacitating. When multiple types of suffering occur together, their individual effect may multiply rather than add to one another. For example, a person suffering from severe depression or anxiety who acquires severe pain from an unknown source may adopt self-blame, creating much more suffering than would result from the sum across individual suffering sources.

3.3 Sex, Age, and Income Differences in Suffering

This section discusses background factors that interrelate with suffering. The specific definitions of these variables are spelled out in Table 3.4 and their descriptive statistics given in Table 3.5.

Research on pain has generally revealed persistent demographic patterns, and some of these tendencies hold true for depression and anxiety. Perhaps the most consistent difference is that women experience more pain than men do. Age produces a more complex pattern in that pain tends to slightly increase with age but sometimes the trend goes downward after age 50.

Table 3.5 Demographic and quality of life indicator estimates of USA adults

Social indicator name	Prevalence in percent of adults	Prevalence in population count	N of cases represented in sample	95 % conf. interval, lower bound	95 % conf. interval, upper bound
Poverty	0.1336	28,131,715	57,260	0.1228	0.1443
Immobility	0.0650	14,904,000	6,231	0.05576	0.0724
Age		Mean age = 46	65,919	45.76	46.91
Age 55+	0.3253	74,662,729	65,919	0.3113	0.3393
Sex (female)	1.52	118,574,569	65,919	1.5	1.53
Income		Mean = 6.18	3.671	6.05	6.31
Qualoflife		Mean = 2.1	24,905	2.07	2.13
Social support		Mean = 2.2	24,864	2.2	2.27
Overall health		Mean = 2.3	65,811	2.24	2.31

Source See Table 3.2

Table 3.6 Suffering indicators by gender

	Male (%)	Female (%)	All (%)
Physical suffering	16.2	21.6	19.0
Extreme pain	2.8	3.8	3.3
Anxiety	8.0	8.7	8.4
Extreme anxiety	3.1	5.8	4.5
Depression	7.4	10.9	9.2
Extreme depression	4.4	6.3	5.4
Grief	2.2	3.4	2.8
Existential suffering	4.5	5.1	4.8
Mental suffering	12.6	15.4	14.0
Poverty	11.1	14.9	13.0
Immobility	5.1	7.7	6.4
Social suffering	14.6	20.1	17.4

Source See Table 3.2

Table 3.7 Suffering indicators by age group

	18–54 %	55+ %	All %
Physical suffering	14.4	29.2	18.9
Extreme Pain	2.6	4.3	3.1
Anxiety	7.7	9.2	8.2
Extreme anxiety	4.7	4.4	4.6
Depression	8.4	10.0	8.9
Extreme depression	4.7	6.3	5.2
Grief	2.1	4.1	2.7
Existential suffering	4.6	4.9	4.7
Mental suffering	13.3	15.0	13.8
Poverty	14.5	9.4	13.0
Immobility	3.4	11.8	6.0
Social suffering	16.6	18.9	17.3

Source See Table 3.2

The gender gaps in suffering are shown in Table 3.6 and difference in suffering between those under age 55 and over 55 are shown in Table 3.7. As found in previous studies, women reported greater suffering than men did, but in this study, more women experienced suffering than men in nearly every type of suffering measured. Furthermore, the gender gap (simple percentage difference) in suffering was relatively greater, the more extreme the suffering. For example, the gender gap for those suffering from anxiety was slightly less than 1 %, but for those suffering from extreme anxiety, it was 2.7 %. The average gap across all suffering indicators was 2.6 %, which does not seem large, but it consistently demonstrated that women had higher suffering on each type of suffering.

In contrast, the gaps in suffering between younger and older age groups are significant, but somewhat inconsistent. While the physical suffering was large (14.8 %) with the older age being higher, the remaining differences in suffering by age group were somewhat smaller. For one indicator, poverty, the gap was in the opposite direction. Specifically, those in poverty were more likely to be found in the younger age groups. Immobility was the reverse, so the two essentially cancel each other out, and social suffering does not show much difference between the two age groups. The average difference between the two age groups across the 12 comparisons in Table 3.7 was only 2.5 %, with higher suffering for the older group except in two instances.

Actually, as you can see in Fig. 3.2, social suffering drops from 20 % to about 12 % for those in their 40s and then rises among those 50 and older. The most striking trend revealed by the chart is the steep, nearly linear relationship between physical suffering and age group: less than 10 % of those in their 20 s had chronic pain or physical suffering, whereas 30 % of those 60 and older reported having had it for the past three months. The remaining types of suffering, including those not shown in the chart revealed no major differences across the age categories.

Regarding income differences by type of suffering, the data reveal a modest correlation between suffering and income. As shown in Fig. 3.3, less and less suffering (either mental or physical) was found among those in higher income groups. In other words, for American adults, the less money earned, the greater the likelihood of either physical or mental suffering.

3.4 Quality of Life and Suffering

The extremely close relationship between various types of suffering and self-reported quality of life stands out from the statistical graph of Fig. 3.4.

Those with a high subjective quality of life had very low levels of all types of suffering, and those with very poor quality of life were relatively very high for all types of suffering. The smooth lines in Fig. 3.4 reveal two important patterns. One is that the relationship between suffering and quality of life follows an exponential curve and second, the curves are steeper at the low end of the quality of life scale, which means that they mutually reinforce each other more at lower levels of quality of life than at higher levels.

Earlier we discovered that while income does not have a large effect on suffering, it does play an important role in quality of life. Income level mediates the role of suffering in shaping quality of life. Those earning over $75,000 annually experience less suffering (either physical or mental) than those earning less, which is depicted in Fig. 3.5.

This finding suggests one possible explanation of the previous finding by Kahneman and Deaton (2010) of happiness increasing with increased income up to the $75,000 annual income threshold, but then it loses its efficacy. Happiness

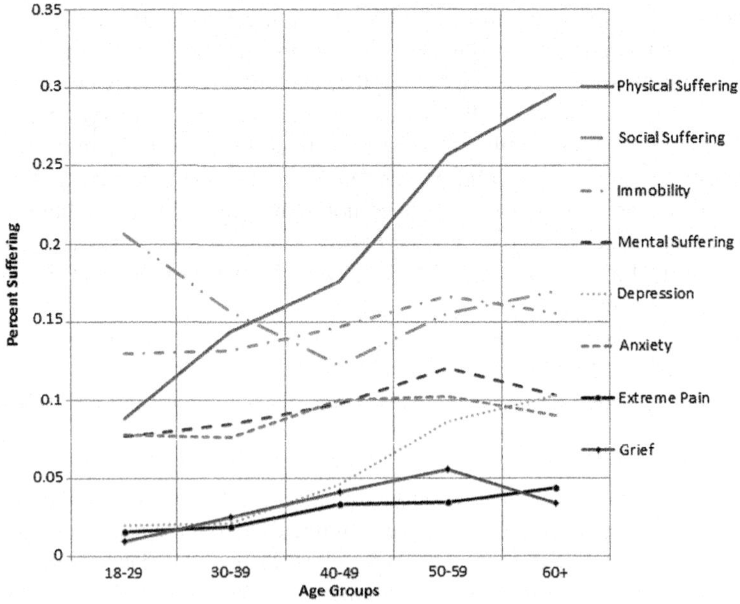

Fig. 3.2 Types of suffering by ascending age groups–the types of suffering in the legend are ordered from largest prevalence to the smallest prevalence for those at age 60+

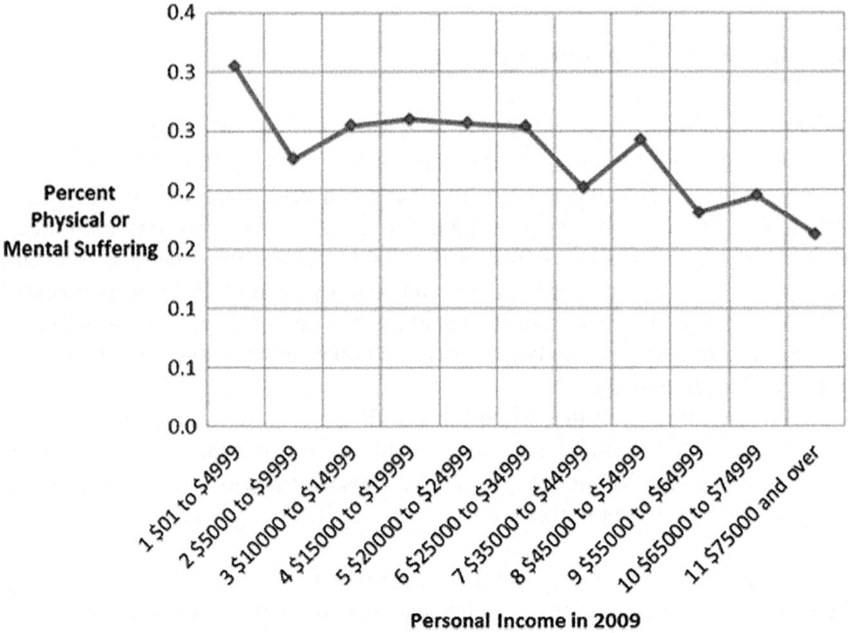

Fig. 3.3 Percent with any mental or physical suffering by income

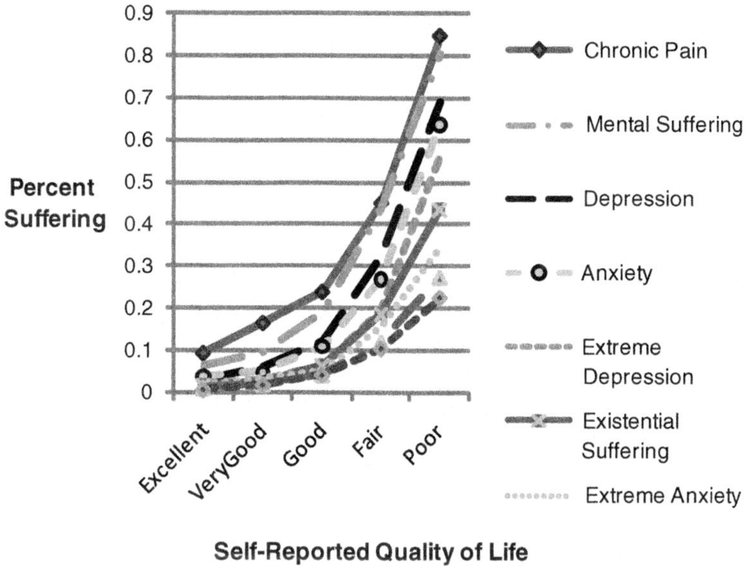

Fig. 3.4 Suffering rises more sharply with each decline in quality of life–The types of suffering in the legend are ordered from largest to smallest rate given in the right-most category on the X-axis, which in this chart is "poor"

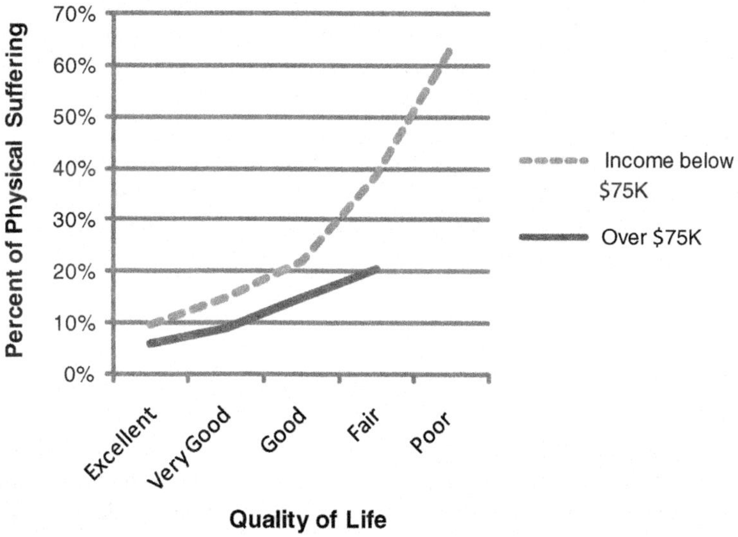

Fig. 3.5 Percent with chronic pain by quality of life and income

Table 3.8 Physical-mental suffering is the best predictor of subjective quality of life

	B	Std. error	Stand. beta	F	Probablity
(Constant)	2.71	0.13		20.55	0.00
Physical-mental	0.19	0.04	0.25	4.95	0.00
Social support	0.13	0.03	0.23	4.79	0.00
Overall health	0.11	0.03	0.21	4.00	0.00
Income	−0.03	0.01	−0.15	−3.25	0.00
Sex	−0.06	0.05	−0.05	−1.21	0.23
Age	0.00	0.00	0.01	0.23	0.82

and quality of life are conceptually different, but empirically highly correlated. It is plausible that the $75,000 income threshold makes income less potent because under the threshold, persons are not able to afford techniques that contain suffering.

3.4.1 Predicting Quality of Life

Clearly, quality of life (QOL) and suffering can be mutually reinforcing or reciprocally influencing each other. However, this analysis so far has tacitly presumed that QOL predicts suffering and background variables like age and income. Background factors of income, overall health (self-reported on a five-point scale), and satisfaction with one's social support all correlate with QOL.

On the basis of the earlier findings, a regression model was applied to the simple model of predicting quality of life (QOL) from overall health, social support, income, sex, age, and Physical-Mental suffering, which is the presence of either physical or mental suffering. The results appear in Table 3.8. All of the predictor variables are significant in predicting QOL except for age and sex. Of the remainder, suffering is the largest contributor to prediction of QOL as measured by the standardized Beta coefficients.

These results should be regarded as exploratory rather than definitive. Additional analytical procedures could be used to verify that this finding did not result from the way the data were modeled. Replications of the analysis in other sets of data are needed as well.

3.5 Implications

The finding that suffering plays a greater role in predicting quality of life (QOL) than overall-health, income, and social support may be one of the most important findings uncovered in this analysis. The other major implication here is that suffering needs to be studied much further, especially in the context of QOL and happiness studies. It is intriguing to learn that suffering may play a greater role in influencing QOL than such factors as poverty, social isolation and overall health.

3.6 Summary

Since this analysis included so many different types of suffering, it seems helpful to recap the findings. First, consider the role of chronic pain, which we equated with physical suffering, acknowledging that it has mental components. An estimate of 19 % of adults with chronic pain is low compared with most studies of chronic pain in western countries, but it seems more valid because so many prior studies used such vague indicators to assess it. In any event, clearly pain (physical suffering), whether it be chronic or extreme, is strongly associated with sex, age, QOL, and to some extent income.

Mental suffering (which was measured as any instance of serious depression, anxiety, grief or existential suffering), had significant but not strong relationships with sex, age, and income. The existential suffering component did not have much of association with anything except QOL. Perhaps, existential suffering dragged down the quality of the mental suffering indicator. Clearly, more thought and work is called on the existential suffering construct. Not surprisingly, the prevalence of grief-related depression was quite low. None-the-less, we found major differences, i.e., women, elderly, and people with low QOL were more likely to suffer from grief.

Social suffering was the most original indicator used and the one most weakened by a lack of internal consistency. Its two elements, poverty and immobility, in some cases cancelled each other out. For example, younger people were more likely to have poverty but older age groups were much more likely to have physical immobility. None-the-less, social suffering was clearly more prominent among women, those over 65 in age, those in lower income brackets, and those with low QOL. Clearly more work is needed on this concept and potential indicators.

3.7 Conclusions

As the science and measurement of suffering has only just begun, it is impossible to precisely know how many people in a given society suffer a particular kind of distress. Just the same, the statistics reported here give us a profile of American adults with minimal sampling error. The challenge is to know exactly what people had in mind when they answered questions in the health survey.

Compared to other surveys of chronic pain around the world, the estimates of suffering prevalences in the IHIS study are unusually conservative. Even so, it is hard to grasp the gravity of the problem when the results tell us that 52 million people, about 27 % of American adults, have a significant case of pain, depression, or anxiety. Finding so much suffering in a contemporary, affluent society raises the possibility that affluence itself through lifestyles and mental frames produces types of suffering not typically found in poverty stricken nations.

Even if we limit our count of the suffering to those who have '*extreme* suffering,' the analysis discovered that 13 % of Americans (25 million adults) struggle with that level of extreme suffering. To some extent, such severe suffering is not preventable, but much of it is. For instance, if poverty were eliminated, nearly 5 million fewer American adults would be subject to extreme suffering. Or if we were to cut US adult obesity in half, as many as 3 million fewer American adults would be distressed by extreme suffering. In other words, at least 20 % fewer adults would suffer from extreme suffering if obesity in the USA were to drop by 50 %. (This estimate presumes that obesity increases serious illness or injury, which in turn results in extreme suffering).

The interesting implication of these conclusions is that a lot of extreme suffering could be prevented by health and social policy. This knowledge offers a new perspective to address an old problem, extreme human suffering. The question is how much do we value the reduction of human suffering and what are we willing to sacrifice so that others may be given an opportunity to escape their suffering?

About half of those who shoulder the distress of any one type of suffering, also suffer from one or more other types. In some cases, they may be causally related. Their inter-relatedness suggests that they all prove burdensome for human beings and in extreme instances, they become incapacitating. When multiple types of suffering occur together, their individual effect may multiply rather than add to one another. For example, a person suffering from severe depression or anxiety who acquires severe pain from an unknown source may feel self-blame, creating much more suffering than would result from the sum across individual suffering sources.

All humans suffer from pain, worry, stress and grief from time to time. Yet not everyone suffers severely, and suffering is distributed unevenly across societies. The word "suffering" has been used in so many ways that researchers have largely neglected its investigation. This exploration begins to bring greater precision to the notion of suffering. It also, suggests how it can be measured and hopefully adds to our understanding of the concept of suffering and its relevance to human well-being and better quality of life.

Human suffering can best be understood from the accumulation of knowledge about the causes, contexts and results of suffering. The breadth and complexity of suffering call for many disciplines including the humanities, social sciences, biological sciences, and health care professionals to compile diverse knowledge bases that can be woven into a deep fabric of understanding.

This exploration begins with the promise that the interplay between suffering and quality of life deserve scrutiny. Major suffering undermines the freedom to live in line with one's own choosing. Major suffering greatly damages, if not destroys one's the quality of life, also called flourishing, thriving, or well-being. In fact, suffering is so intertwined with quality of life that it may be useful to treat suffering as an indicator of a negative quality of life. It even appears useful to conceptualize suffering as not only a component but also a cause and outcome of quality of life.

Measurement of all types of suffering except pain has been given very little systematic attention. No researcher has attempted to use the concepts of mental and social suffering in survey methodology. This pioneering work will benefit greatly from future work on these challenges.

References

American academy of pain medicine (2013). Facts and Figures on Pain. American Academy of Pain Medicine. http://www.painmed.org/PatientCenter/Facts_on_Pain.aspx. Accessed 20 May 2013.

Borsook, D. (2012). A future without chronic pain: neuroscience and clinical research. http://www.dana.org/news/cerebrum/detail.aspx?id=39160 Accessed 8 March 2013.

Bourdieu, P., et al. (2000). Understanding. In P. Bourdieu, et al. (Eds.), *The weight of the world: social suffering in contemporary society*. Stanford: Stanford University Press.

Breivik, H., Collett, B., Ventafridda, V., Cohen, R., & Gallacher, D. (2006). Survey of chronic pain in Europe: prevalence, impact on daily life, and treatment. *European Journal of Pain, 20*(4), 287–333.

Cassell, E. J. (2004). *The nature of suffering and the goals of medicine*. USA: Oxford University Press.

Elzahaf, R. A., Tashani, O. A., Unsworth, B. A., & Johnson, M. I. (2012). The prevalence of chronic pain with an analysis of countries with a human development index less than 0.9: a systematic review without meta-analysis. *Current Medical Research and Opinion, 28*(7), 1221–1229.

Hardt, J., Jacobsen, C., Goldberg, J., Nickel, R., & Buchwald, D. (2008). Prevalence of chronic pain in a representative sample in the United States. *Pain Medicine, 9*(7), 803–812.

Harstall, C., Ospina, M. (2003). How prevalent is chronic pain? *pain clinical updates*. Seattle, WA: International association for the study of pain *11*(2), 1–4.

Institute of Medicine Report. (2011). *Relieving pain in america, a blueprint for transforming prevention, care, education and research*. Washington: The National Academies in Press.

Johannes, C. B., Le, T. K., Zhou, X., Johnston, J. A., & Dworkinin, R. H. (2010). The prevalence of chronic pain in United States adults: Results of an internet-based survey. *Journal of Pain., 11*(11), 1230–1239.

Johnson, P. J., Blewett, L. A., Ruggles, S., Davern, M. E., & King, M. L. (2008). Four decades of population health data: The integrated health interview series as an epidemiologic resource. *Epidemiology, 19*(6), 872–875. doi:10.1097/EDE.0b013e318187a7c5.

Kahneman, D., Deaton, A. (2010). High income improves evaluation of life but not emotional well-being. *PNAS* 107(38), 16489-16493. http://wws.princeton.edu/news/Income_Happiness/Happiness_Money_Report.pdf. Accessed 7 March 2013.

Kleinman, A., Das, V., & Lock, M. (Eds.). (1997). *Social suffering*. Berkeley: University of California Press.

Land, K. C., Michalos, A. C., & Sirgy, M. J. (2012). The development and evolution of research on social indicators and quality of life. *Land and others (Ed.), Handbook of Social Indicators and Quality of Life Research* (pp. 1–22). New York: Springer.

Langle, A. (2008). Suffering—an existential challenge: Understanding, dealing and copy with suffering from an existential-analytic perspective. International *Journal of Existential Psychology & Psychotherapy, 2*(1), 39–50. http://journal.existentialpsychology.org/index.php?journal=ExPsy&page=article&op=view&path%5B%5D=115&path%5B%5D=58. Accessed 18 May 2013.

Macdonald, G. M., & Jensen-Campbell, L. A. (Eds.). (2010). *Social pain: Neuropsychological and health implications of loss and exclusion*. Washington: American Psychological Association.

Tsang, A., Von, K., M, Lee, S., Alonso, J., Karam, E., et.al. (2008). Common chronic pain conditions in developed and developing countries: Gender and age differences and comorbidity with depression-anxiety disorders. *The Journal of Pain. 9*(10), 883–891.

Wilkinson, I. (2005). *Suffering: A sociological introduction*. Cambridges, UK: Polity Press.

Williams, B.R. (2004). Dying young, dying poor: A sociological examination of existential suffering among low-socioeconomic status patients. *Journal of Palliative Medicine, 7*(1), 27–37.

Chapter 4
Suffering on a Global Scale

This analysis compares countries using maps, charts and tables to convey the seriousness of recent statistics pertaining to suffering across the globe. In an age of 24 h media coverage, the *idea* of global suffering is easy to comprehend. But it is still challenging to estimate precisely how many, or what percent, of people face one or more major calamities or social traumas in any given year. This is to say, we know that there is great suffering in the world, but we have not gotten our arms around how much and where. Another hurdle is reached because no one has tried to rigorously estimate misery or suffering until now. Global suffering measures are uncharted territory.

4.1 Applying the Notion of Global Suffering

In 1988, a group primarily concerned with containing the rate of world population growth created an "International Human Suffering Index" (Kelley 1989) using data from the World Bank. The Index was a composite of the following statistics for each of nearly 100 countries: GDP per capita, average inflation rate, average growth of urban areas, infant mortality rates, access to clean water, energy consumption, and a rating of the extent of human freedom in each country. The researchers did not produce specific estimates of suffering by country, but highlighted how the population growth rate in the 30 countries lowest on their Index was between 2 and 4.2 %. The implication was that countries with such high annual growth rates would double in population every 20–30 years and population pressures were likely to increase suffering.

There have also been numerous and wide-ranging international comparison studies of well-being since 1989, but none has claimed to estimate human suffering. The only cross-national studies producing information pertaining to suffering have been health studies of chronic pain or depression. In 2003, Breivik et al. (2006) conducted a phone survey of 46,000 adults residing in 14 European countries (plus Israel) to evaluate chronic pain. While the prevalence of chronic pain

R. E. Anderson, *Human Suffering and Quality of Life*, SpringerBriefs in Well-Being and Quality of Life Research, DOI: 10.1007/978-94-007-7669-2_4, © The Author(s) 2014

lasting over 6 months averaged 19 %, that range included Spain, where 12 % of respondents reported suffering chronic pain, and Norway, with 30 %. Over two-thirds of those suffering chronic pain were receiving medical treatment, and half were receiving medications. Most chronic pain sufferers were less able to work and one-fifth had lost their latest job.

Another large chronic pain study, Tsang et al. (2008), included samples from 10 developed nations and 6 developing nations, plus the cities of Beijing and Shanghai. Overall, this team's estimate of the prevalence of chronic pain was 38 %. While the rates of chronic pain in developing and developed countries revealed no overall difference, when the statistics were standardized by age, chronic pain prevalence was slightly higher in developing nations (41 %) than developed nations (37 %). Tsang et al. also found that an average of 10 % of all populations reported depression/anxiety disorders. Another noteworthy finding was that those with chronic pain were over twice as likely to report depression/anxiety disorder compared to those without chronic pain.

In recent literature, Bromet et al. (2011) conducted a survey of 90,000 adults in 18 countries to assess major depression as defined by the DSM-IV.[1] They contrasted 10 highly developed countries with 8 developing countries and found a prevalence rate of 15 % for developed countries versus 10 % for developing countries for major depression during the previous 3 months. Only about 3 % reported their depression had lasted over 12 months. The study found that those suffering from major depression reported a high likelihood of social and employment impairment. Women were much more subject to depression than men were; however, unlike the pattern for chronic pain, those in the younger and middle age groups had a higher prevalence of depression than those over age 65.

4.2 Subjective Suffering

While studies of life satisfaction and quality of life have not claimed to measure suffering, it has become apparent that they sometimes produce useful information about suffering. Studies comparing nations or regions in terms of well-being and quality of life use two types of measures. One is subjective 'life satisfaction' such as the Cantril Ladder instrument described below. The second consists of collecting official statistics and building a composite index or indicator. This is the approach taken by the UNDP (2010) Human Development Index and its variants.

Life satisfaction, sometimes called well-being or subjective well-being, is based upon respondents' evaluation of their lives as a whole. The timeframe for this evaluation may be the present time, the past 5 years, the next 5 years, or all three. Researchers generally assume that this well-being is a unitary concept, but

[1] Diagnostic and Statistical Manual of Mental Disorders, 4th Edition.

some have pointed out that the negative end of the continuum may not be a simple absence of positive well-being, but, rather, ill-being (Headey, Holmstrom and Wearing (1984). As 'ill-being' is not a colloquial word, this label has not caught on. Regardless of its label, this chapter will show how focusing on the negative end of the life satisfaction continuum can provide information on suffering as well.

The Cantril Self-Anchoring Striving Scale (Cantril 1965) has been included in several Gallup research initiatives, including Gallup's World Poll of 150 countries and Gallup's in-depth daily poll of America's well-being (Gallup-Healthways Well-Being Index; Rath and Harter 2010). The Cantril Scale measures the well-being continuum representing judgments of life or life evaluation (Diener, Kahneman, Tov and Arora 2009). In one application of the Cantril Scale across several countries, Deaton (2008) found substantial correlation between the Cantril Scale and income. The effect of income on satisfaction, however, drops off after an annual income point of $75,000 USD—that is, money may aid happiness and life satisfaction, but only to a point (Kahneman and Deaton 2010).

The Cantril Self-Anchoring Scale is typically administered by an interviewer with the following instructions:

> Imagine a ladder with steps numbered from zero at the bottom to 10 at the top. The top of the ladder represents the best possible life for you and the bottom of the ladder represents the worst possible life for you. On which step of the ladder would you say you personally feel you stand at this time? (Show visual of a ladder.) On which step do you think you will stand about 5 years from now? (Rath and Harter 2010)

The Gallup adaptation of the Cantril Ladder includes a question on the present and one on future life satisfaction. For this analysis, the Gallup World Poll indicator of present life satisfaction was compared with data published in the United Nations Human Development Report (2010). After eliminating those countries lacking life satisfaction data, my resulting dataset contained statistical data for 123 countries. The total world population in mid-2010 was estimated at 6,852,000, and the population of the 123 countries analyzed in this study total 6,596,000 million. Over 96.5 % of the actual world population in 2010 is included in the dataset.

I produced my 'subjective suffering' indicator from the formula $11 - X$, where X is a national average of life satisfaction from the Gallup polls. For the purposes of creating a world map (Fig. 4.1), I truncated the subjective suffering values to whole numbers and combined the two highest numbers as there were only four countries in the highest category. This process yielded five categories or levels. The highest level of suffering is represented by Level 5, while the lowest level of suffering is Level 1.

In Fig. 4.1, the darker the shading, the more intense or severe the subjective suffering. Stark white areas indicate missing data (e.g., Greenland, Paraguay, and Angola). Level 5, which represents the highest suffering, is represented by the black seen in a number of central African countries, plus Bulgaria, Haiti, and Afghanistan. Level 4 nations include South Africa, Turkey, and India; Level 3 includes Egypt, China, and Chile; Level 2 includes Argentina, the UK, and Japan; and

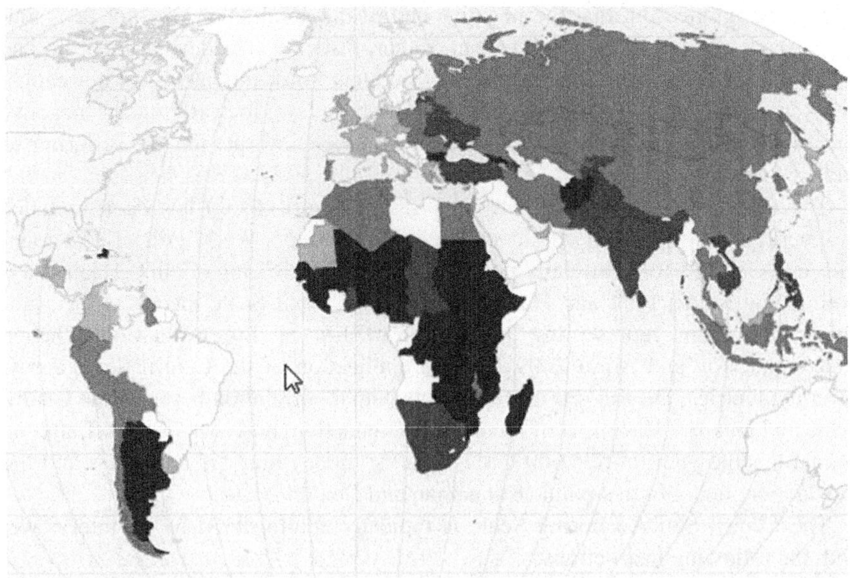

Fig. 4.1 Subjective suffering levels worldwide (123 countries) See Table 4.1 for more detail

Level 1 includes the USA, Saudi Arabia, and Brazil. All of these country rankings can be seen in Table 4.1, where countries appear in their rank from highest to lowest subjective suffering. The right hand column of Table 4.1 lists the countries for which no data were available.

4.3 Types of Calamities Related to Suffering

In the past century, the enormous network of public and private organizations providing international humanitarian aid and development assistance has built a system of collecting and publishing statistical data regarding human progress and human problems. These data make it possible to evaluate the effectiveness of aid programs more and more accurately. Most of the data are aggregated country by country, yielding estimates of demography, health, and dozens of different types of calamities and social traumatic events, such as infant mortality, deaths due to cancer, homicides, suicides, prevalence of malnutrition, and so forth. Arguably, the most expansive and well-known international statistical agency is the United Nations Development Program (UNDP), which has been publishing its annual Human Development Report (HDR) for 20 years.

In an attempt to create a more objective measure of suffering, this substudy combines those indicators reported by the UNDP in their 2010 Human

Table 4.1 Countries at each level of subjective suffering from level 5 (highest subjective suffering) to level 1 (lowest subjective suffering)

Level 5 highest	Level 4	Level 3	Level 2	Level 1 lowest suffering	No information
Tanzania	Central AR[a]	Algeria	Belize	Brazil	Albania
Togo	Botswana	Iran	Kuwait	Dominican Rep.	Angola
Zimbabwe	Ghana	Estonia	Malaysia	Spain	Bosnia
Burundi	Sri Lanka	Indonesia	El Salvador	Mexico	Brunei
Benin	Cambodia	Mongolia	Jamaica	Saudi Arabia	Croatia
Liberia	South Africa	Hungary	Singapore	Panama	Gibbon
Burkina Faso	Armenia	Egypt	Italy	Venezuela	Greenland
Congo (DR)	Kyrgyzstan	Morocco	Uruguay	New Zealand	Iraq
Sierra Leone	Tajikistan	Lithuania	Japan	Austria	Ivory Coast
Kenya	Namibia	Slovakia	Greece	Netherlands	Jordan
Madagascar	Azerbaijan	Tunisia	Czech Rep	Australia	Laos
Mali	Bangladesh	Peru	Honduras	Sweden	Lebanon
Mozambique	Nepal	Portugal	Trinidad[b]	United States	Libya
Niger	Ukraine	Romania	Mauritania	Canada	Myanmar
Nigeria	Chad	Russia	Argentina	Finland	New Guinea
Cameroon	Pakistan	Hong Kong	Nicaragua	Switzerland	North Korea
Haiti	Viet Nam	Uzbekistan	Cyprus	Ireland	Oman
Afghanistan	Latvia	Kazakhstan	Israel	Norway	Paraguay
Ethiopia	India	Malawi	France	Denmark	Serbia
Rwanda	Philippines	Chile	Slovenia	Costa Rica	Syria
Angola	Turkey	Korea (So.)	Guatemala		Turkmenistan
Zambia	Belarus	Thailand	Germany		Yemen
Georgia		Ecuador	Colombia		
Bulgaria		China	UAR		
Guinea		Bolivia	Belgium		
Senegal		Guyana	United Kingdom		
Sudan		Poland			
Uganda					

Within each column of Table 4.1, the countries are listed from highest suffering to lowest suffering

[a] CAR represents Central African Republic

[b] Trinidad represents Trinidad and Tobago

Development Report. (The 2011 and 2012 HDR reports have been released, but they include fewer indicators than the 2010 report.) Most indicators of social trauma (like war fatalities) can be reported either as population counts or prevalences (the proportion of the population) experiencing the type of suffering or calamity. I used prevalences in this sub-study to construct a composite measure of objective suffering. Twelve of these calamity types are listed in Table 4.2.

Note that the first 10 types of calamity in Table 4.2 pertain to physical suffering, while the last two represent social suffering as defined in Chap. 1. By totaling

Table 4.2 Prevalence of types of global calamities from the 169 country-based statistics in the Human Development Report (UNDP 2010), in which all prevalences represent 2009 or latest available year

Global calamity types	Prevalences
Hunger (nutrition deprived)	857,915,108
HIV prevalence	33,446,568
Internally displaced persons	25,297,883
Refugees (outflow)	12,757,786
Infant deaths (before-age-5)	10,530,830
Pollution-related deaths	5,030,203
Disasters (deaths and homeless)	3,381,851
Suicides	576,850
Homicides	302,093
Civil war deaths	103,437
Total physical suffering	949,342,609
Poverty (multidimensional)	1,587,702,000
Corruption (# reporting bribes)	668,986,000
Total social suffering	2,256,688,000

across the types of physical calamities, we get an estimated global population of nearly one billion who physically suffered during 2009. This is a rough estimate that needs additional work, but it is staggering none-the-less.

The social suffering included in the last two categories includes poverty (affecting about 1.5 billion persons) and corruption (adding another 700 million). Country poverty estimates are based upon the UNDP (2010) indicator of 'multidimensional poverty', which includes measures of health and education as well as living standards.

The statistics in Table 4.2 represent total populations or counts of calamity victims, whereas for the correlational analyses, these totals or counts are converted to rates or percentages, so that the indicators are not contaminated by variation in population sizes across countries. What follows in the remainder of this section is a summary description of each of the 12 calamity types.

4.3.1 Hunger (Nutrition Deprived)

The UNDP data tables include estimates of hunger, the intensity of food deprivation best described as protein-energy malnutrition. The reported estimate used here is the average percent of the population suffering malnutrition due to a "shortfall in minimum dietary energy requirement". In other words, this statistic gives us an estimate of the share of the population whose daily food intake was below their dietary required minimum energy level. This form of hunger leads to serious health problems and early death.

4.3.2 HIV Prevalence

HIV prevalence estimates are typically reported as the number of persons aged 15–49 living with HIV. To obtain a relative measure of HIV for each population, the prevalence counts were multiplied by an age group's proportion of the population. Across our focal countries, the average HIV prevalence was 1.6 %. The sum of HIV prevalence across countries was about 34 million–nearly identical to estimates by the World Health Organization.

4.3.3 Internally Displaced Persons

The UNDP data provide an estimate by country of the number of Internally Displaced Persons (IDP) having fled their homes while remaining within the country.

4.3.4 Refugees (Outflow)

The UNDP data provide an estimate by country of the number of refugees who fled from any given country to another. The data came from the UN Refugee Agency (UNHCR) estimates of total refugees under their responsibility or that of the UN Palestine relief agency. This totals 12.7 million including refugees, asylum-seekers, returnees, internally displaced, and stateless persons. It does not include those who are still in asylum-seeking (pending) status or internally displaced persons.

4.3.5 Infant Deaths (Under-age-5)

This measure is similar to infant mortality, except that it includes all deaths of children before their fifth birthday. This statistic is usually reported as deaths per 1,000 live births. To calculate the number of child deaths per country, I adjusted for fertility rates and total population to estimate the total under-age-five child deaths per year as a percentage of the total population.

4.3.6 Pollution-related Deaths

This includes known deaths officially attributed to pollution. These deaths include those due to unsanitary water and air pollution and include such deadly conditions as lung and cardiovascular diseases. It is almost a given that this indicator underestimates the actual number of deaths due to pollution, both because some reporting systems remain primitive and because the determination of cause of death may be hindered in the absence of advanced technology and professional skills.

4.3.7 Disaster Victims

I constructed this estimate with help from the World Health Organization's (WHO) International Disaster Database (EM-DAT). The estimate of victims includes deaths and displaced persons to both natural disasters and man-made (or anthropogenic) disasters.

4.3.8 Suicides

The WHO also provided the statistics used on annual suicides. Such estimates are only available for about 80 countries, so this indicator has more missing data points than other categories. Like homicide, suicide is generally considered a crime from a legal standpoint, but as a crime against self, it reflects a very specific form of suffering.

4.3.9 Homicides

Homicides are typically reported per 100,000 persons. Using each country's population, I calculated the percent of intentional homicides in the total population for each country. The data came from the UNODC (United Nations Office on Drugs and Crime) 2010 report.

4.3.10 Civil War Deaths

The UNDP Report estimates fatalities from civil war by country, based upon the average of years of conflict year between 1990 and 2008. The estimates used here are deaths per million persons. I calculated the total fatalities by multiplying these relative estimates by the population in millions.

4.3.11 Poverty

This poverty indicator is called 'multidimensional poverty' in the HDR 2010 report. This measure has three major dimensions: health, education, and living standards. Health and education have two indicators each, but living standards has six: assets, housing floor, electricity, water, toilet, and cooking fuel. Although the poverty measure is not based upon a minimum income level, the number of eligible people by these criteria is roughly the same number as those who live on less than $1.50 USD per day.

4.3.12 Corruption

While the corruption indicator is contained in the HDR 2010 report, its source was the Gallup World Poll database. The Gallup polls asked the sample of each country if they had "faced a bribe situation this past year". The country level data was simply the percent who answered 'yes'.

4.4 Construction of an Objective Suffering Indicator

One of the challenges of combining estimates of different types of suffering is that they naturally overlap. For instance, many of those persons who suffer malnutrition also live in poverty. In this analysis, I addressed the overlap problem by generating statistical estimates of overlaps between pairs of variables and adjusting the joint prevalences for each pair. The grand total of estimated world suffering is about three billion people, or 44 % of the world population in 2009. Without adjustment for overlapping categories, the estimate would have been about four *billion* people.

In terms of more accurately estimating—or refining measures of—global suffering, future work should go in two directions. One would limit the estimates of suffering to *severe* suffering such as painful, chronic illnesses and premature deaths. Another direction might attempt to be more inclusive, considering, for instance, prevalences of domestic abuse and rape. Of course, this is limited by the absence of consistent standards of reporting across nations. In time, such improvements in objective suffering may arise.

Consider now the challenge of evaluating the validity of the estimates of 12 types of calamities as components of an indicator of objective suffering. To evaluate the predictive validity of each calamity that might help compose a total measure of objective suffering, each calamity type was correlated with subjective suffering and with income as measured by Gross National Income per capita. Table 4.3 reveals the results. More of the calamity types were correlated with subjective suffering than with income, surprising given that life satisfaction and income have been found to be closely correlated in other studies.

The estimates of suffering in Table 4.2 tend to underestimate global suffering. They do not include failed states such as Somalia and Iraq, which did not have stable governments, primarily because of long-term civil wars. Such countries are

Table 4.3 Disposition of the correlations of the 12 calamity types

Most correlated with subjective suffering	Most correlated with income (GNI per capita)	Not correlated with either
Hunger	Disaster casualties	Corruption
Infant deaths	Homicides	Pollution deaths
HIV prevalence		Displaced persons
Poverty		Refugees
Civil war deaths*		

*Civil War Deaths was not used in the objective suffering score because it did not contribute to the prediction of subjective suffering above what other calamity types had already contributed

Table 4.4 Means, standard deviations, and correlations for modeling of calamity types (N = 122)

	Means	SDs	Suffering	HIV	Child deaths	Poverty
Suffering	5.1	1.5	1.00			
HIV prevalence	0.02	0.02	0.34	1.00		
Infant deaths	35.	36.0	0.72	0.28	1.00	
Poverty	26.3	27.6	0.66	0.25	0.56	1.0
Hunger	14.9	0.02	0.63	0.30	0.28	0.7

Table 4.5 Linear regression predicting subjective suffering from four calamity types

Model	B	SE	Standard B (Beta)	Significance
1. Constant	4.06	0.13		0.00
2. HIV Prevalence	6.40	3.85	0.16	0.03
3. Infant deaths	0.02	0.00	0.48	0.00
4. Poverty	0.00	0.00	0.18	0.01
5. Hunger	0.18	0.005	0.21	0.04

Note: R-square = 0.58; N = 122 countries

excluded because it was impossible to survey a sample of adults at that time. If these nations were to be added to the existing 123 countries used in this sub-study, the levels of suffering would certainly rise.

The top four calamity types in the first column of Table 4.3 were used as components of a composite indicator of objective suffering, and a linear regression model was used as a basis for refining the weights of these components. I present the results of the model in Tables 4.4 and 4.5. These four variables explain 58 % of the variance in the subjective suffering indicator—not bad for four very heterogeneous calamity types. Figure 4.2 visualizes the strong predictive relationship of the model described in Table 4.5. The strong linear relationship between the four weighted composite calamity predictors (X-axis) of this model and subjective suffering adds credibility to claims of validity for both subjective and objective suffering.

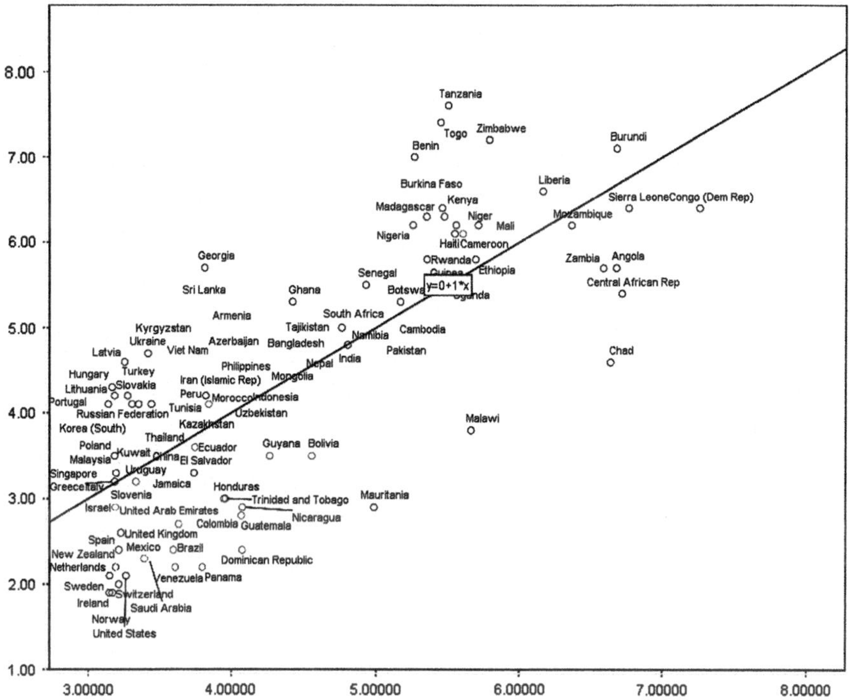

Fig. 4.2 Scatterplot of Subjective Suffering on the vertical Y-axis predicted by the objective suffering composite of four factors (hunger, poverty, child deaths, and HIV prevalence) distributed on the horizontal X-axis

4.5 Multidimensional Suffering and Gender Inequality

Since the validity of the objective suffering measure has been established, the subjective and objective measures can be combined to yield a more robust social indicator of total suffering. Thus, I adjusted the objective and subjective scores for their ranges and added them together to produce the multidimensional suffering indicator (MSI).

To explore how useful this multidimensional suffering indicator (MSI) might be, I now turn to the relationship between MSI and gender inequality. (Later, I consider system social support). The role of gender inequality in suffering has not been widely understood in most development and political circles. Notably, however, Martha Nussbaum (2001a) has helped articulate how unequal treatment of women, especially within developing societies, undercuts development initiatives. By making it extremely difficult for girls and women to contribute their capabilities to productive work, including family and community decision-making, many societies with high gender inequality develop slowly and erratically. Nussbaum's "capabilities approach" for development calls for eliminating violence, health disadvantages, education deficits, and other disparities that keep women, racial minorities, and other social groups from applying their potential toward progress and contributing toward the reduction of suffering (Nussbaum 2001a, b, 2011; Nussbaum and Sen 1993).

For my analysis, I use the multidimensional measure of gender inequality developed by the UNDP and used in the HDR (2010) report. It consists of the following components: (1) maternal mortality ratio, (2) adolescent fertility rate, and (3) the share of parliamentary seats held by each nation. The UNDP combined these three sub-indicators into a single variable by calculating the geometric mean of each of the three indicators for each gender and then by combining them statistically (HDR 2010, p. 219).

The resulting 'gender inequality' indicator reflects the loss in human development resulting from women's disadvantage in reproductive health, empowerment, and the labor market. Country scores range from 0 (complete gender equality) to 1 (worst possible women's advantage).

The fact that higher gender inequality appears as a significant statistical predictor of higher multidimensional suffering in this analysis suggests that gender inequality is a significant cultural barrier to human well-being and the reduction of suffering.

Figure 4.3 depicts the regression of multidimensional suffering on gender inequalities for the 123 countries in the 2010 HDI dataset. Because of the rounded slope, a quadratic equation fit the data much better than a linear model. The resulting R^2 was 0.5, indicating that half of the variation in suffering was accounted for by gender inequality alone.

The shape of the scatterplot distribution illustrates that as gender inequality initially increases, suffering increases very little. Once a substantial degree of gender inequality exists, however, suffering begins to rise sharply. This curve results from such outliers as Saudi Arabia, which has little suffering but is very high on gender inequality, on the right extreme; and Burundi, with very high suffering and moderately high gender inequality, on the left. Removing these two outliers would straighten the curve somewhat, but because there are quite a few countries with substantial gender inequality but only modest levels of suffering, the overall shape of the relationship between the two variables would change only slightly.

I notice with interest that, between 2001 and 2010, a handful of countries (including the United Arab Emirates, Kuwait, and Albania) moved from an extremely high level of gender inequality to a moderate level of gender inequality. This shows that, with government leadership, it is possible to instigate major change in a society's culture of gender inequality, and to do so in a relatively short period.

Although not shown here, gender inequality has a very strong association with development as measured by the UNDP's Human Development Index (HDI). While one could argue that the HDI drives gender equality, the stronger causal influence appears to be that gender equality promotes human development. The rising levels of education among women and the increasing norm of gender equality in terms of personal well-being results in a more productive work force, and in turn, a rising standard of living. The close connection between gender equality and the reduction of suffering, as shown in Fig. 4.3, seems to be a positive demonstration of that larger link between gender equality and human progress.

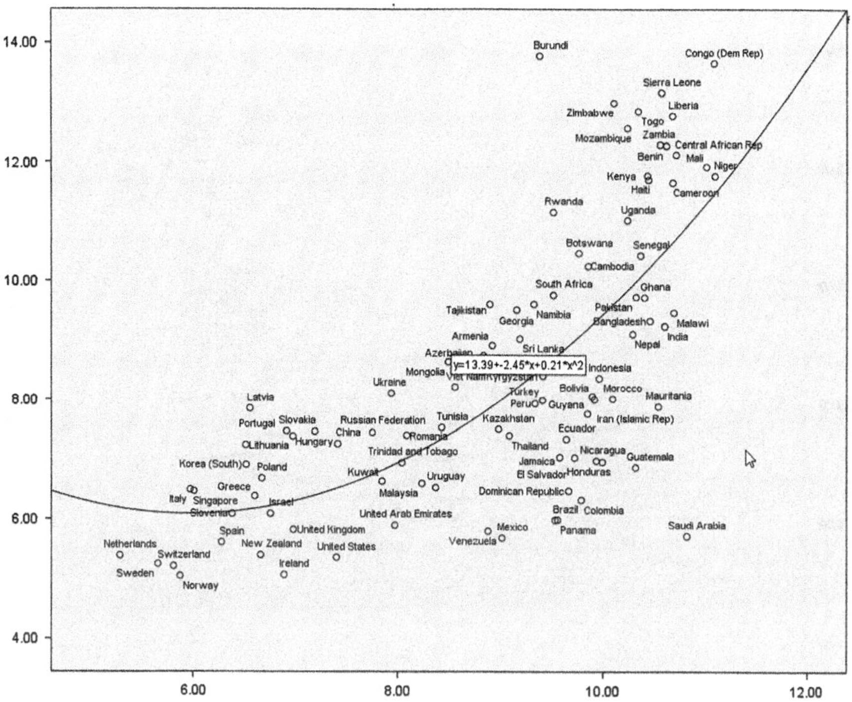

Fig. 4.3 Scatterplot of multidimensional suffering (*vertical* Y-axis) predicted by 'gender inequality' (*horizontal* X-axis) revealing a strong curvilinear relationship in which as gender inequality initially increases, suffering does not increase until the moderately high end of the gender inequality continuum

4.6 Multidimensional Suffering and Social Supports

Next, I look to the role of 'social support' systems in reducing suffering. Figure 4.4 depicts the moderately strong, linear relationship between the two. 'Social support', distributed horizontally in the chart, is based upon a question asked in the Gallup World Poll: "Do you have a social support network"? The indicator is the percent in the country who answer 'yes'. As seen in Fig. 4.4, the percent affirming a social support network ranged from 25 to 100 % of the nations in this study.

The countries with the most people claiming a social support network were Venezuela, Canada, Australia, New Zealand, Switzerland, Finland, and Netherlands at 94 %. Among the top fifth of the nations were the United States, Brazil, and Costa Rica. The countries with the fewest adults (< 50 %) admitting to a social support network were Togo, Burundi, Benin, and Pakistan.

One important consequence of graphically depicting suffering throughout this chapter is that the scatterplots show that all of the countries at the high

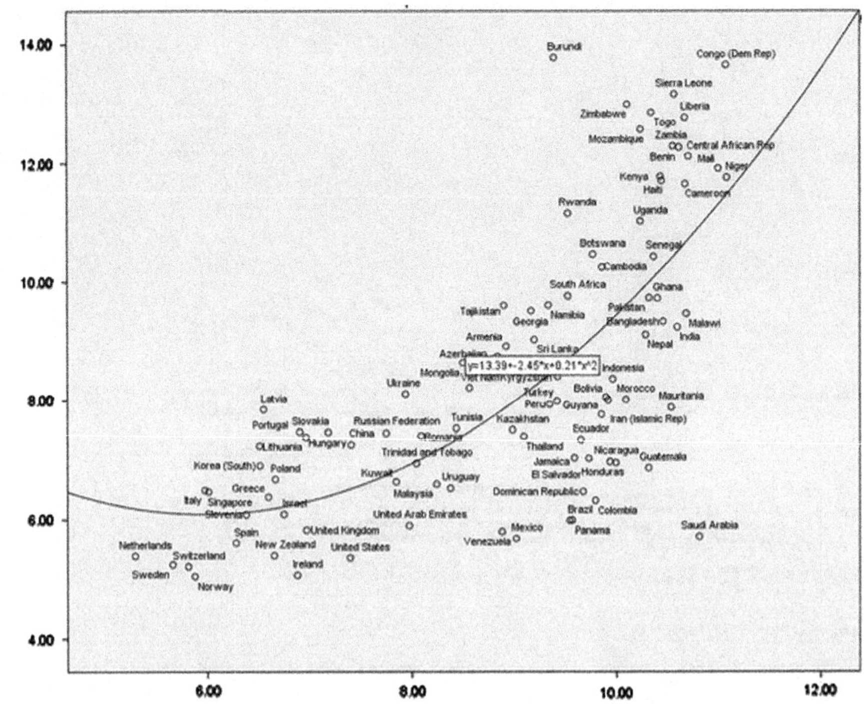

Fig. 4.4 Scatterplot of multidimensional suffering (*vertical* Y-axis) predicted by level of 'social support' (*horizontal* X-axis) revealing a strong linear relationship in which greater 'social support' yields lesser suffering ($R^2 = 0.5$)

end of the suffering continuum are African, except for Haiti, Pakistan, Georgia, and Cambodia. Clearly, the dominant source of severe national suffering lies in Africa.

As mentioned earlier, several 'failed states' were not included in this analysis because they were too dangerous or fractious to survey. The following countries, which are rated high on the Failed States Index, were not included: Somalia, Sudan, Iraq, Myanmar, North Korea, Yemen, Libya, and Iran. These eight nations have a combined population of 276.8 million, or slightly less than 5 % of the world population. If these nations are combined with the Level 5 suffering states, the combined Level 5 population is over 16 % of the world's total population, slightly over one billion people. This number is similar to Collier's (2007) estimate of the world's population most seriously trapped by poverty.

The implicit presumption of Fig. 4.4 is that suffering may be, at least in part, a consequence of whether the people of any given country have (or believe they have) an adequate social support system. While it is plausible that a great deal of suffering is moderated by a strong social support system, it may also be that shared suffering can decay and even destroy a social support system. Such a breakdown of social

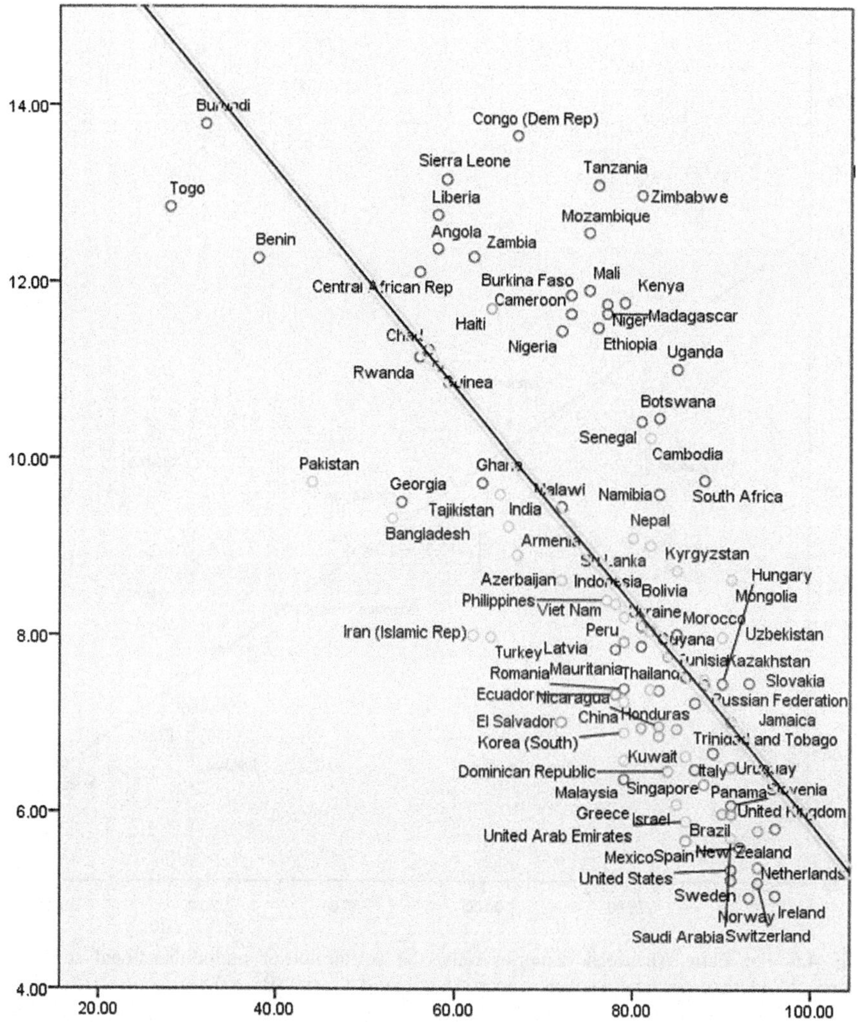

Fig. 4.5 For Europe and North America only, a scatterplot of suffering (*vertical* Y-axis) predicted by 'social support' (*horizontal* X-axis) ($R^2 = 0.5$)

support might easily result from a major civil war or a catastrophic disaster. In some instances, less severe calamities may also erode the social support networks.

To explore the role of region in the pattern of these relationships, I examined the same correlations and regression model for just the 19 European and North American countries. The results are presented in Table 4.5. The patterns in this subset of more affluent nations follow the same structure as the entire set of 123 countries. Eastern European countries lie in the upper-left corner, high on suffering and low on social support. Western European countries and the USA rest in

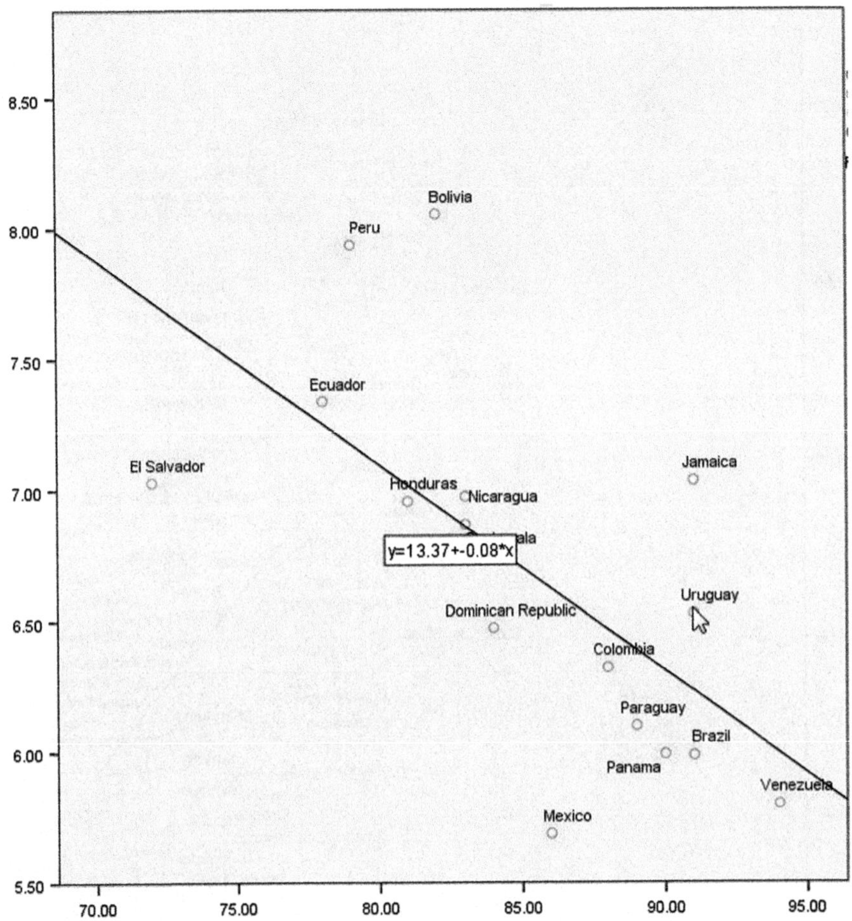

Fig. 4.6 For Latin American countries only, the scatterplot of multidimensional suffering (*vertical* Y-axis) predicted by 'social support' (*horizontal* X-axis) ($R^2 = 0.43$)

the lower-right corner, low on suffering and with 90 % or more of their people reporting having a social support network. The principal exception to this pattern is Greece, an extreme outlier with slightly higher suffering than other Western European countries on suffering, but a much lower percentage of its people reporting a 'social support network'.

An equivalent pattern can be seen in Fig. 4.6, which contains only Latin American countries. Unlike the distinct separation of Eastern and Western nations in Fig. 4.5, the two extreme subgroups of Latin American nations do not have much in common. The two subgroups probably interact with one or more variables that might explain their distinctness; hopefully, additional research will add insight into the nature of the subgroup pattern in Fig. 4.6.

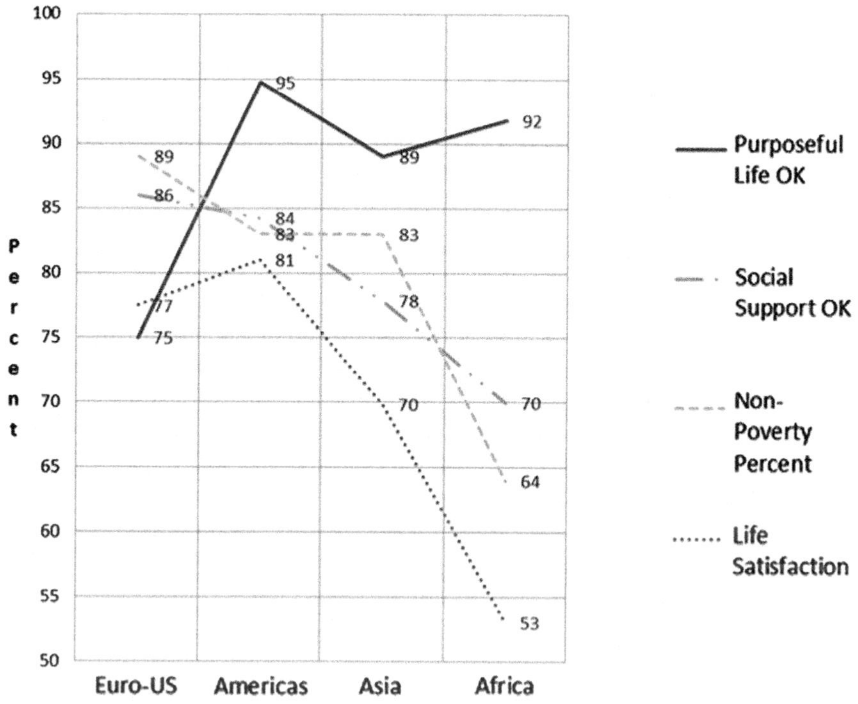

Fig. 4.7 Percent with 'purposeful life', social support network', non-poverty population, and 'life satisfaction' for each of four major regions: Europe including the USA, Latin America, Asia, and Africa. using one-way analysis of variance, the differences in percentages on each of the four lines are statistically significant

To better understand the role of social support networks, particularly within Latin Americans nations, I display the mean percentages for several indicators for each of the four most populous world regions graphically in Fig. 4.7. Each line represents a social indicator, and the four data points on each line correspond to the region at the bottom of each grid column.

The bottom line depicts life satisfaction averaged by region. We see that Latin America had the highest life satisfaction, followed by Europe, Asia, and finally Africa. Note that this pattern is somewhat similar to the two lines above, which represent 'social support networks' and the percent of the population not in poverty. These trends lines roughly be the same were we to have plotted either average income or the level of development as measured by the Human Development Index.

The one indicator that departs wildly from that pattern is the top line, which plots the percent of people in each region who say they have a 'purposeful life.' Latin Americans, and a few African countries, were much more likely to report a purposeful life than were people in other regions. Latin America also is the region with the highest life satisfaction, which illustrates one of the biggest puzzles in happiness research: Latin American countries have much lower income and development levels as compared to European countries, but their citizens report a higher than expected happiness and contentment with life.

The most intriguing aspect of Latin America's reported sense of purposefulness is whether it might explain why Latin Americans are so happy and satisfied with life. If indeed, Latin Americans are more likely as individuals to have a sense of meaningful purpose in life, then this probably translates into an optimistic outlook on life. Optimism is sure to raise self-reported levels of well-being. Research to explore what makes life so meaningful for Latin Americans, or at least the majority of their populations, would be quite illuminating. For instance, Latin Americans tend to be socialized to place very high value on family and community, and research might explore whether this socialization accounts for their unusually high sense of purpose. It also seems likely that Latin Americans acquire a sense of meaning or purpose gained from the solidarity and trust felt in their families, religion, and communities.

4.7 Conclusions

Until now, quantitative measurement of social suffering as an attribute of social systems has not been attempted. Quantitative research on suffering at the individual level has been neglected as well. Numerous empirical studies have included pain measurement at both physiological and subjective levels (Cassell 2004; Nordgren et al. 2011), but these studies generally do not link pain to suffering.

This chapter's analysis is the first to begin to quantify the distribution of suffering around the world. Now it is possible to make preliminary comparisons regarding differences in degrees of suffering across countries and regions. Armed with this knowledge, suffering can be taken into account in public policy considerations. For instance, if one nation contemplates occupying another nation, raising the potential for millions of people being displaced or killed, estimates of such suffering should be calculated and weighed against any potential benefits of entering into military conflict.

Clearly, careful reconsideration is urgently needed on policy agendas for reduction of suffering in failed states and other nations with extreme suffering. The challenges are enormous. Environmental sustainability, political and economic stability, ethnic and social integration, preparedness for disasters, healthcare, and population control are all paramount in preventing suffering. A combination of resources and an international peace corps may make major inroads in the reduction of suffering, unless a climate of violence creates a spiral of disintegration.

There is another caveat I must raise: this study examines suffering and related variables at one point in time. Longitudinal data and analysis are needed to chart *progress* in the reduction of suffering. This applies to communities and towns, as well as nations and the entire global community. Conversely, while comparative national suffering gives us unique and useful insights, the examination of smaller aggregates may be even more useful. For example, provinces, cities, or communities can be tracked over time, and targeted efforts can help alleviate small scale suffering before it becomes entrenched.

Finally, one key finding from cross-national comparative analysis of subjective suffering is that social support networks play a role in diminishing suffering. Yet few aid organizations have policies directed toward building social support systems or enhancing social cohesion in developing countries.

So far, focus of this book has been on suffering as an outcome largely of social dysfunctions like poverty and various types of calamities. However, as Chabal (2009) points out, suffering can serve as the *precipitating* cause of poverty, violence, and other health risks. Future research should address this challenge with reciprocal causation models.

References

Breivik, H., Collett, B., Ventafridda, V., Cohen, R., & Gallacher, D. (2006). Survey of chronic pain in Europe: prevalence, impact on daily life, and treatment. *European Journal of Pain, 20*, 287–333.

Bromet, E., et al. (2011). Cross-national epidemiology of DSM-IV major depressive episode. *BMC Medicine 9*:90. http://www.biomedcentral.com/1741-7015/9/90/. Accessed on 25 April 2013.

Cantril, H. (1965). *The pattern of human concerns*. New Brunswick: Rutgers University Press.

Cassell, E. J. (2004). *The nature of suffering and the goals of medicine*. Oxford: Oxford University Press.

Chabal, P. (2009). *Africa: The politics of suffering and smiling*. London: Zed Books.

Collier, P. (2007). *The bottom billion—Why the poorest countries are failing, and what can be done about it*. New York: Oxford University Press.

Diener, E., Kahneman, D., Tov, W., & Arora, R. (2009). Income, aging, health and wellbeing around the world: Evidence from the Gallup World Poll. In E. Diener (Ed.), *Assessing Wellbeing* (pp. 233–246). Oxford: Springer.

Headey, B., Holmström, E., & Wearing, A. (1984). Well-being and ill-being: different dimensions? *Social Indicators Research, 14*(2), 13–115.

HDR (2010) *Human Development Report 2010*. See UNDP (2010).

Kahneman, D., & Deaton, A. (2010). High income improves evaluation of life but not emotional well-being. *PNAS 107*(38), 16489–16493. http://wws.princeton.edu/news/Income_Happiness/Happiness_Money_Report.pdf. Accessed on 13 April 2013.

Kelley, A.C. (1989) The international human suffering index: reconsideration of the evidence. *Population and Development Review 15*(4), pp. 731–737.

Nordgren, L. F., Banas, K., & MacDonald, G. (2011). Empathy gaps for social pain: why people underestimate the pain of social suffering. *Journal of Personality and Social Psychology, 100*(1), 120–128.

Nussbaum, M. (2001a). *Upheavals of thought: The intelligence of emotions*. Cambridge: Cambridge University Press.

Nussbaum, M. (2001b). *Women and human development: The capabilities approach*. Cambridge: Cambridge University Press.

Nussbaum, M. (2011). *Creating capabilities: The human development approach*, Harvard University Press, cambridge.

Nussbaum, M., & Sen, A. (1993). *The quality of life*. Oxford: Clarendon Press.

Rath, T., & Harter, J. (2010). *Well-Being—The first essential elements*. New York: The Gallup Press.

Tsang A, Von Korff, M, Lee S., Alonso, J. Karam, E. et al. (2008). Common chronic pain conditions in developed and developing countries: gender and age differences and comorbidity with depression-anxiety disorders. *The Journal of Pain. 9*(10), 883–891.

UNDP (2010) *Human Development Report 2010.* New York, NY: Palgrave Macmillan for the United Nations Development Program.

Chapter 5
World Suffering Expands as Gaps in Care Widen

The previous two chapters have uncovered the vast reaches of suffering across the globe in both affluent and developing societies. Given that the amount of suffering is so enormous, it cannot be taken lightly. In this chapter, we review some of the steps that individuals and institutions can take to address the many challenges of suffering.

5.1 Alternative Approaches of Responding to Suffering

Suffering comes from three places: individuals, institutions, and forces of nature. Suffering that arises strictly from 'acts of God,' like earthquakes and tornadoes, does not fall within our primary concern because it is largely unpreventable. Suffering is preventable when it follows in part from human choices such as driving too fast and playing with a loaded gun. With disasters such as hurricanes, we do not know to what extent the ensuing suffering is preventable because hurricanes now gain their force from both natural and man-made forces.

From the data presented on global suffering in earlier chapters, it is evident that most suffering, especially in non-affluent countries, results from illness, injuries, disability, and poverty, especially poverty-related scarcities. The scarcities are forces largely under human control through social institutions. Strategies for relieving suffering deserve careful attention, because it may be possible to reduce suffering through improved healthcare policy, such as public health programs (Fancher 2003; Farmer 1997; Kleinman 2011).

The source of the suffering provides clues for how the suffering can be alleviated or prevented. From the standpoint of taking action, this discussion uses the language of 'relief' to discuss suffering in the present, and the vocabulary of 'prevention' for suffering that might occur in the future. Using this distinction between relief and prevention, Table 5.1 distinguishes individual relief actions from institutional ones while contrasting the three major types of suffering: physical, mental, and social, which were discussed in the Chap. 1.

R. E. Anderson, *Human Suffering and Quality of Life*, SpringerBriefs in Well-Being and Quality of Life Research, DOI: 10.1007/978-94-007-7669-2_5, © The Author(s) 2014

Table 5.1 Approaches to reducing suffering for various types of suffering

	Physical suffering	Mental suffering	Social suffering
Relieving suffering (individual level)	• Empathy, compassion, caring and social support • Assistance in balanced living, stress management, and meditation • *Physical therapy* • *CAM*treatments* • *Non-narcotic and narcotic pain medication* • *Professional counseling and pain management*	• Empathy, compassion, caring and other social support • Assistance in dissociation from negative emotions such as anger; gratitude and acceptance of others • Cultivating presence, acceptance, and help for others • *Treatments including hypnosis as needed* • See "Physical suffering for more tactics	• Empathy, compassion, caring & other social support • Food, nutrition assistance • Access to water, sanitation, fertile land as needed for self-preservation • Health support for illnesses and injuries • Ending violence & structural violence, war & civil conflicts
Preventing suffering (institutional level)	• Long-term institutional changes • Education and training in above techniques • Training in meditation, contemplation & self-compassion	• Long-term institutional changes • Training in ending intolerance and discrimination • Training in social responsibility, forgiveness, care ethics, reconciliation, • Training in meditation, contemplation and self-compassion	• Long-term institutional changes • Training in human rights including health & education, access to water and sanitation • Education in human values, nonviolence, reconciliation

Note Items in italics, largely, require professional services; *CAM* Complementary and alternative medicine

Note that the first item in all of the columns of the table has "empathy, compassion, caring and social support" as a form of remedial help that individuals can offer no matter the type of suffering. The table reveals that social suffering in general requires support and relief at the institutional level from government agencies and other human service organizations, because the roots of social suffering are embedded into the fabric of communities and societies.

Mental and physical sufferings, on the other hand, are more subject to the unique characteristics and experiences of individuals. Consequently, each individual may require a unique regime of treatment and other social supports. Various therapies and medications, including self-therapy and self-medication, apply to both individual and institutional approaches to relief. The columns of 'Physical' and 'Mental' suffering in Table 5.1 suggest examples of tactics for suffering alleviation. These examples offer steps that you can take to ameliorate the suffering of others as well as your own.

Institutional support systems may include any kind of organization: governmental or non-governmental, formal or informal, local or global, family or non-family that provides relief in response to suffering. In some instances, the institutional response is directed to a grouping of people who share the same plight.

A specialty of medicine and healthcare, called pain medicine, has emerged to respond to physical suffering or pain of different types (Cassell 2004). Likewise, the professions of psychiatry, psychology, and other related treatment communities cater to needs for relief from mental suffering. These approaches traditionally respond to the suffering of individuals.

Reducing social suffering, in contrast, requires organized or institutional responses because of its highly intertwined linkages with social contexts such as communities. Relief of social suffering may require major change in these social environments. Most of the institutional change needed falls under the banner of social justice (Wronka 2008), which includes human rights and development standards, as well as legal statues. Not only does the relief of suffering require radical expansion of global health programs, but education in non-violence with training in conflict resolution and reconciliation are needed as well. (See Table 5.1 for more specifics.)

Although not all prevention effort need be focused on a long-term time frame, the prevention section of Table 5.1 does emphasize education and training and such activities that imply the prevention of suffering require a long time. This is necessary because the social institutions that perpetuate current suffering are deeply embedded in the fabric of global society and the power structures that maintain it. Both segregation, based upon race/ethnicity, wealth, religion, and any other status characteristics, and gross inequality pose a nearly overwhelming challenge to the eradication of needless suffering.

While professional healthcare is likely to have the greatest chance of alleviating instances of suffering, informal help and supportiveness may also relieve the hurt and distress. Likewise, nonprofessional healthcare workers, both paid and unpaid, may ease suffering as well.

All major religions and most ethical traditions define helping others who suffer as a moral obligation, unless there are special circumstances such as the likelihood of causing even greater harm by helping (Armstrong 2011). Helping behavior may be facilitated by the moral emotions of empathy, care, and compassion (Gilbert 2009).

Despite the moral imperative to come to the rescue of anyone unjustly suffering, discourse on the topic has become problematic by the imprecision, ambiguity and multiple meanings of the words available to discuss this fundamental aspect of social life. The semantic state of the words compassion, caring, caregiving, altruism, sympathy and compassionate action have been called "conceptual chaos" by McGaghie et al. (2002). Because the terminology is imprecise, measurement of this phenomenon has been impeded.

Table 5.2 Quality of suffering relief: dimensions and states

Dimensions	Minimal	Optimal
Depth	Superficial	Root causes removed
Breadth	Single symptom	Multiple symptom
Elapse time of relief	Temporary	Permanent
Generality	Single sufferer	Many relieved
Completeness	Partial recovery	Complete
Affordability	Excessive	Affordable for all

In order to construct measures of compassion and willingness to help, researchers ask questions regarding persons' recent giving to charity, unpaid volunteer work, and giving of assistance to strangers. While this does not, by any means, fully represent their disposition to be compassionate or to help those enduring severe suffering, it does capture a desire, or a small slice of actual past activity, to care for the needs of others. Furthermore, calculating the percent of people giving such responses and then comparing percentages across countries and regions gives us some insights into cultural tendencies and differences relevant to concern for others. In the discussion of results from cross-national studies below, this concern is described in terms of charity, volunteering, and when multiple behaviors are involved, 'compassionate caring.'

An obvious consideration in approaches to relieve suffering is the underlying effectiveness and quality of any procedure used. Despite the importance of this concept of the quality of suffering relief, it appears to be neglected in the research on quality of life as well as pain management.

As Table 5.2 suggests, it is possible to identify important facets of the quality of suffering relief. These would include depth, breadth, elapsed time, generality, completeness and affordability. The table also includes illustrative low and high states for each of these dimensions. This type of analysis is the first step in the construction of one or more indicators that can help to refine the measurement of the concept. These dimensions may ultimately help us construct measures of the degree of effectiveness of strategies and operations intended to relieve different types of suffering.

5.2 The Care Divide

How well does the world system work to bring relief to those who suffer, especially those who suffer the most? To answer this question, several charts will be shown that reveal the patterns of association between suffering and attempts to relieve suffering for both nations and smaller areas.

Unfortunately, the world order is the opposite of a system designed to maximize social support for those who suffer. Ideally, places with higher levels of suffering would have proportionately more people that cared about suffering and were willing to help those who seriously suffer. This would apply to social grouping as well as physical locations. Here we apply the model to nations, and then rich nations, and finally the states of the United States.

5.2.1 Nations

Generally, the only indicators widely available to assess peoples' compassionate action to relieve suffering are two rather crude measures related to charity, specifically: (1) the percent of residents who report having given any amount to charity during a given time, usually a year and (2) the percent who report having volunteered for unpaid work for charitable organizations. There is an important exception. It comes from the Gallup World Poll asking people the question: "Have you helped a stranger within the last month?" (English 2011).

Look first at Fig. 5.1, which graphically depicts the association between subjective suffering and the percent giving money to charity across 123 countries. The obvious pattern that stands out is a negative relationship between suffering and giving by country. Generally, those societies with the most suffering are the least likely to have people that have given to charity. For example, Tanzania has the most suffering but less than 20 % gave to charity. At the other extreme are Switzerland and Denmark with very little suffering but about three-fourths of their people gave money to charity.

The curved shape of the scatterplot in Fig. 5.1, which is sometimes called a decay curve, is a result of some countries with moderate giving levels also having low levels of suffering. For example, Costa Rica was average in percent giving to charity but was very low on subjective suffering. Such patterns may be a consequence of other factors such as inequality that can cause a relationship to curve like the line in Fig. 5.1. Another way of interpreting the curvilinear relationship is that as the percent giving to charity approaches zero, the upward jump in suffering increases, compared to those nations with high concentrations of charitable givers. Apparently, having very few charitable people in a country also means having an unusually high amount of suffering.

It is well established that there is a very strong relationship between a nation's average income and the amount of giving (CAF 2006; 2012). People in wealthier nations in general are more likely to give to charity than those in less wealthy nations and overall the total amount given in wealthy nations is larger (CAF 2006). To the extent that suffering is correlated with income, the relationship between suffering and charitable giving by nations as shown in Fig. 5.1 is a reflection of the fact that people in richer nations are more likely to give than people in poor ones.

However, the scatterplot tells us much more. It reveals that few people in poor countries where suffering abounds give to charity, perhaps in large part because many are preoccupied with subsistence, getting enough food, and staying

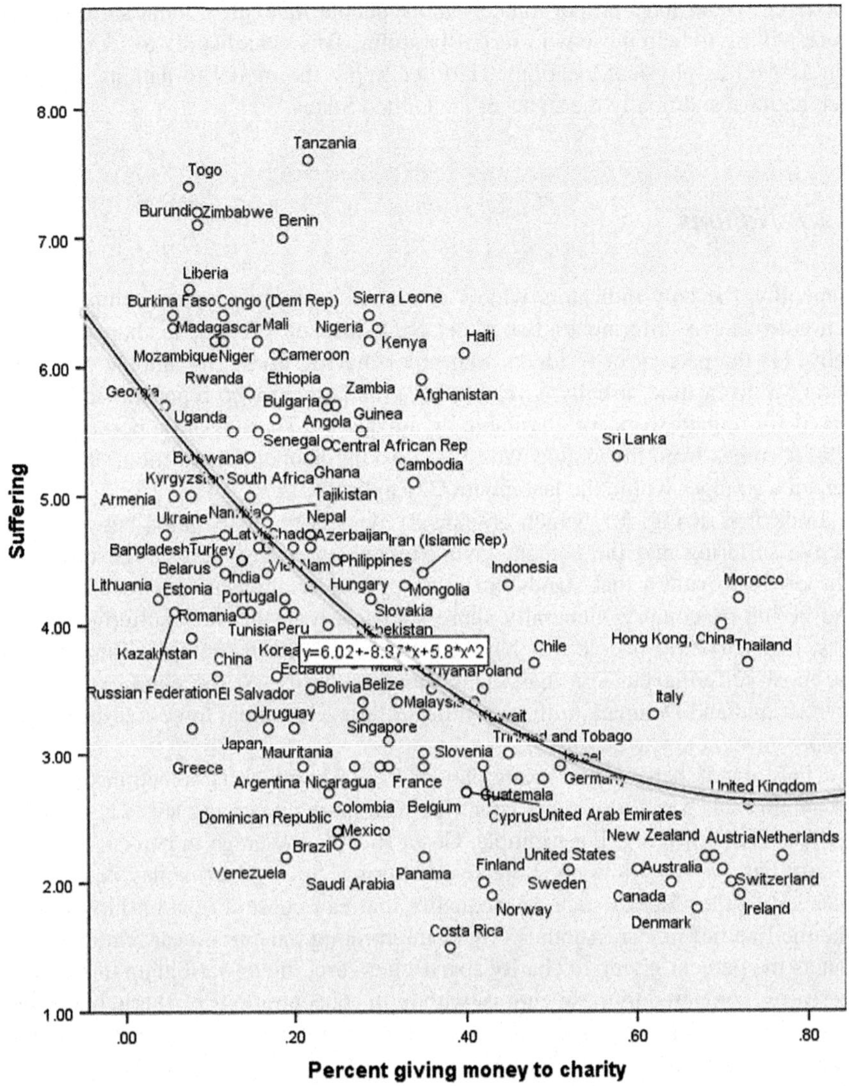

Fig. 5.1 Subjective suffering (*vertical* Y-axis) correlated with the percent of adults who gave to charity in 2009 (*horizontal* X-axis) for 123 countries from the UNDP (2010) Human Development Report (Quadratic $R^2 = 0.36$)

trauma-free from day to day. In contrast, many in affluent countries tend to give to charity because they have more than enough wealth to live comfortable lives.

To the extent that charity is motivated by compassion, this portrayal of the distribution of suffering and charity reveals that the alignment of suffering and compassion may be exactly opposite of what would be ideal for purposes of reducing suffering (and poverty) in the world.

Giving to charity and the amount given are not the only measures of caring and the desire to reduce suffering. Volunteering of one's personal time for relief projects or activities that reduce or prevent suffering also expresses desire to alleviate suffering. Whether people volunteer is largely dependent upon whether organizations are created with a mission to provide a charitable service and then organized to recruit and supervise volunteers. Such organizations are quite common in some affluent countries, but much less so among poor countries. So, volunteering is not always a good indicator of whether or not people desire to engage in charitable activities.

5.2.2 Rich Nations

The OECD (Organization for Economic Co-operation and Development) came up with a way to build an improved index of charitable action by combining giving with volunteering and helping strangers. The OCED called this index 'prosocial behavior,' however, here it will be called 'compassionate caring.' The indicator is based upon survey respondents answering yes to any of the following three questions: (1) Have you done any volunteering work in the past month? (2) Have you donated any money to a charity in the past month, and (3) Have you helped a stranger in the last month? (OECD 2011).

The scattering of rich countries in Fig. 5.2 shows the correlation between suffering and compassionate caring and it is surprisingly similar to that of the 123 countries in the previous scatterplot, Fig. 5.1. ("Rich countries" in this analysis are the 20 nations with the highest Gross National Income (GNI) per capita, excluding those countries like Luxemburg and Singapore that have less than three million population.) With a sampling of countries around the world, both rich and poor, it was understandable to see misalignment between suffering and charity. Finding a misalignment between the amount of caregiving reported by wealthy populations and the average suffering of these populations is less expected.

Wealth, particularly if it is combined with considerable income equality, tends to reduce poverty, which reduces suffering. Furthermore, compassionate caring tends to be more highly correlated with wealth at the national level, because without at least moderate resources, people do not have the means to give much to charity nor spend much time in unpaid work. This would explain such a strong negative relationship between suffering and compassionate caring.

The disturbing feature of this statistical pattern is that it poses a strangle hold on reduction of suffering from compassionate caring, because compassionate caring occurs in societies that need it the least. Countries with considerable subjective suffering, on the other hand, do not have much of a culture of compassionate caring that might provide aid to those who suffer. Those in a state of extreme suffering need caring and help the most, but according to the pattern shown in Figs. 5.1 and 5.2, they are more likely to reside in a community or society that lacks a culture of compassionate caring.

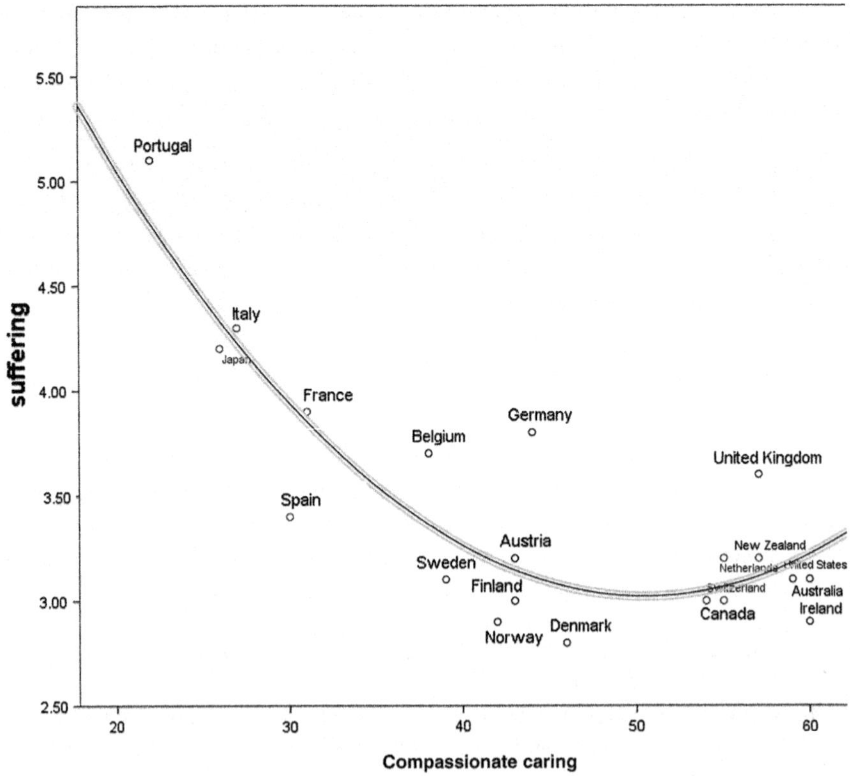

Fig. 5.2 Subjective suffering (*vertical* Y-axis) in relation to compassionate caring (*horizontal* X-axis) for the 20 most affluent nations (Quadratic $R^2 = 0.76$)

5.2.3 *States*

Now, consider the same patterns of suffering but instead of nations, we switch to the 50 US states plus the District of Columbia (Washington, DC). Figure 5.3 shows the relationship between the average hours of volunteering for charitable causes by the poverty rate, which is the percent of each state with an annual income below the official poverty line. Figure 5.4 shows the nearly identical distribution except that the vertical axis (Y-axis) is suffering rather than poverty.

The overall scatter of states is remarkably similar to the scatter in Fig. 5.1 of nations' percent who gave to charity by subjective suffering. Likewise, the relationship between charity and poverty follows the same pattern found among nations.

States that are the highest in both poverty and suffering include Arkansas, Alabama, Louisiana, Mississippi, and West Virginia. Kentucky stands apart in that, while in high poverty, it is moderate in its level of charitable giving. At the other extreme are states with relatively little suffering and poverty. Examples of such states are Hawaii, Alaska, Vermont, Wyoming, and Washington State. Note how all of these states, except for Hawaii, simultaneously have a high percentage of persons who

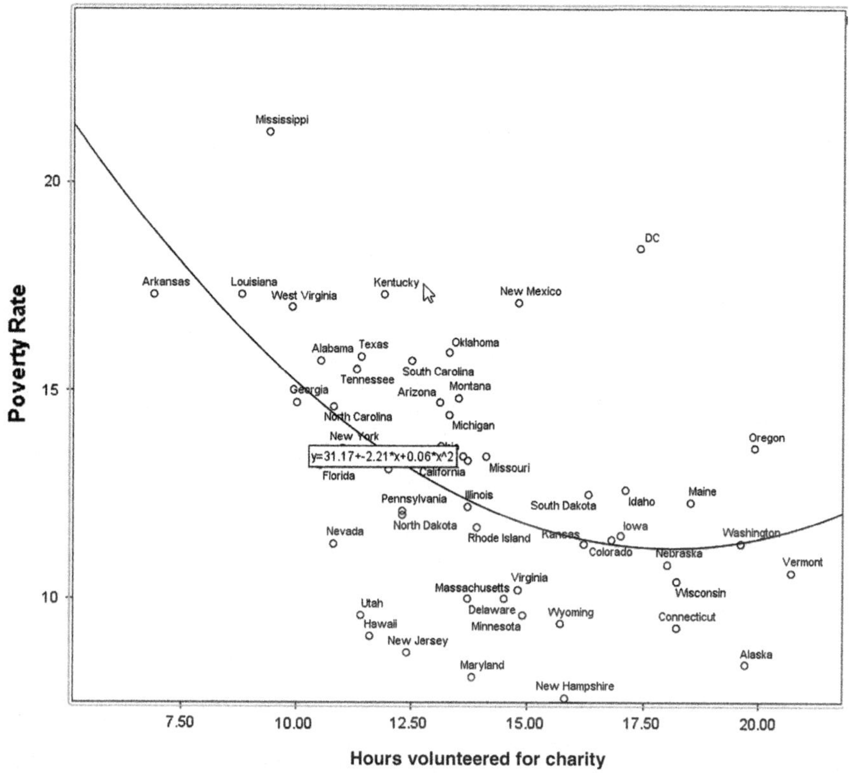

Fig. 5.3 Poverty lines (*vertical* Y-axis) as an effect of average hours volunteered for charitable causes (*horizontal* X-axis) in 50 states plus Wash. DC (Quadratic $R^2 = 0.36$)

gave to charity. Hawaii is a unique state in that it has lower suffering but only a moderate level of giving to charity. Hawaii has a less demanding environment and a more leisurely culture, which may account for its very low level of subjective suffering.

Relatively few people in poor states where suffering is concentrated report giving to charity, perhaps in large part because many are busy with earning enough to pay for food, health and housing from day to day. In contrast, many in the more wealthy states tend to give to charity partly because they have more than enough wealth to live comfortable lives.

Just like in cross-national comparisons, these state by state comparisons portray a misalignment of suffering and compassion, making the relief of suffering much less likely than if compassionate caring and suffering tended to co-occur. Income-based segregation, both across and within countries, tends to concentrate people into enclaves of the rich and the poor, just as cities tend to evolve into ghettoes and gated communities unless income desegregation policies and put into place.

This income-based segregation here is labeled a ‘caring gap’ or ‘caring divide,’ which is both similar and obviously different from the ‘digital divide. Caring divides are pervasive in that they occur not only in neighborhoods, towns, and

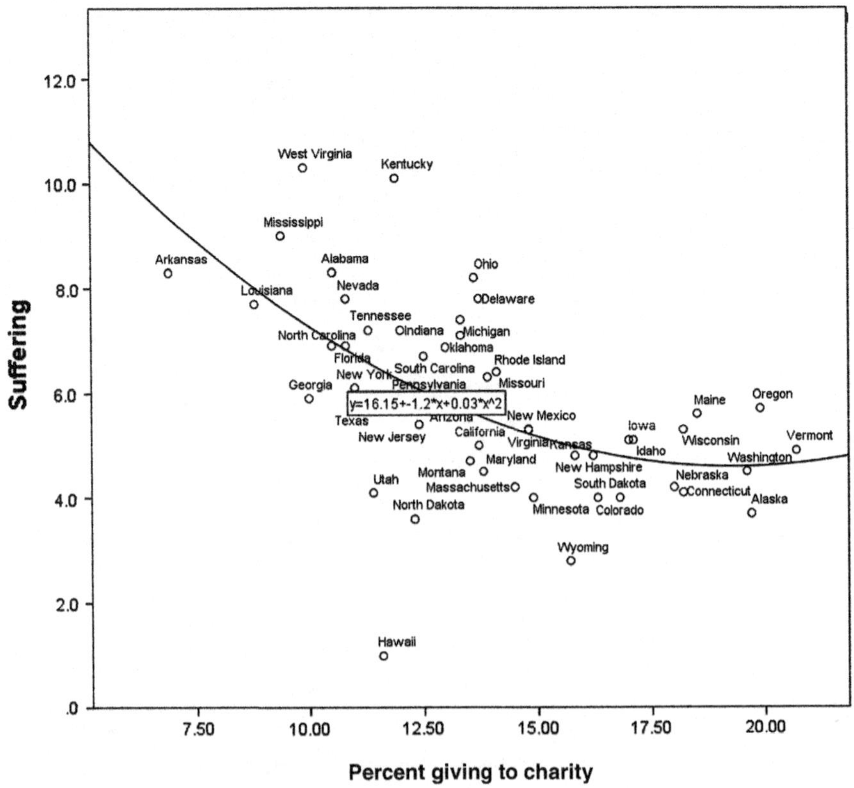

Fig. 5.4 Subjective Suffering (*vertical* Y-axis) as an effect of percent giving to charity (*horizontal* X-axis) in US States (Quadratic R2 = 0.33)

cities, but as visualized here, caring divides characterize the concentration of rich and poor nations as well as the rich and poor states in the USA.

It is not just that the rich and poor are geographically separated, e.g., the global north versus the global south, but within more affluent regions are more caregivers, even though these persons with greater caring capacity are needed more in non-affluent regions.

The caring gap is mostly a product of social inequality. If income and cultural differences between the rich and poor were negligible, the two groups would live side by side, which would close the caring divide.

Bringing people with resources together with those who need help is primarily a matter of community networks and humanitarian institutions, which is often called 'social capital.' Just as the essence of social capital is valued networks of social and institutional relationships, 'caring capital' refers to those networked social relationships involving the giving of care by one party to another without any explicit expectation of reciprocation or reward (Anderson 2012). Caring capital typically involves diverse types of informal giving of care not largely dependent upon formal exchanges of goods or services. For this reason, caring capital tends to be described in words like compassion, generosity, kindness, altruism, charity, and humanitarianism (Boltanski 1993; Cohen 2001; Coleman 1988; Gilbert 2009; Glenn 2000, 2010; Johansson, et al. 2012; Nussbaum 1996, 2001; Salvati 2008; Sorokin 1950; Sznaider 2001).

The relationship between caring and various forms of capital has scarcely been noticed by social scientists, either theoretically or empirically. Major exceptions include the empirical work of Wuthnow (1991) and the theoretical writing of Oliner (2008). The essence of the state of knowledge on this subject is that nations, and states within nations, lack much opportunity for caring capital to function, ultimately leading to the increase, rather than the reduction, of poverty and suffering. Recent research (Taylor and Fry 2012; Gennetian et al. 2013) on increasing segregation has documented a substantial growth rate, especially in large cities, for residential segregation by income, which brings with it greater caring divides.

5.3 How the Rich Undermine Reduction of World Suffering

The closeness of income and the amount of giving is moderated by an additional factor, which some might call greed or stinginess of 'haves' versus the 'have-nots.' Economic studies for some time have found a tendency for the wealthier to give proportionately less, in general than the poor and low income (Frank 1999, 2007; Independent Sector 2002; James and Sharpe 2007). That is, if you calculate the percent of one's after-tax income that is given to charity, the share (percent) of income given for charity declines the more income one receives, with the exception of a few with relatively very high income, e.g., over $150,000 per year.

Charitable giving as a percent of income tends to decline as income increases except for the very high income bracket. Figure 5.5, shows this general pattern: the very poor donate almost twice their share of after-tax income than do the middle income groups, but as the household income rises up to $150,000 per year and up, the proportion given away starts to rise slightly. This rise in proportionate giving among the wealthiest is not stable, in part because the number of the super-rich is still a relatively small population.

Recent studies, including laboratory experiments, found that those who identify with 'lower class' people, as determined by lower income and education, were more likely to be generous in donating to those in a state of suffering or need (Piff et. al. 2010). They found that both family social standing as well as personal identification with a given social strata, affected generosity and desire to reduce suffering (Stellar et al. 2012).

The researchers who conducted these studies under the leadership of Keltner (2009) have mapped several processes by which social strata associated with income and wealth affect compassionate caring for those who are suffering or otherwise need help (Kraus et al. 2010; Kraus et al. 2011). They have found that even though people with relatively fewer resources feel a reduced sense of control over

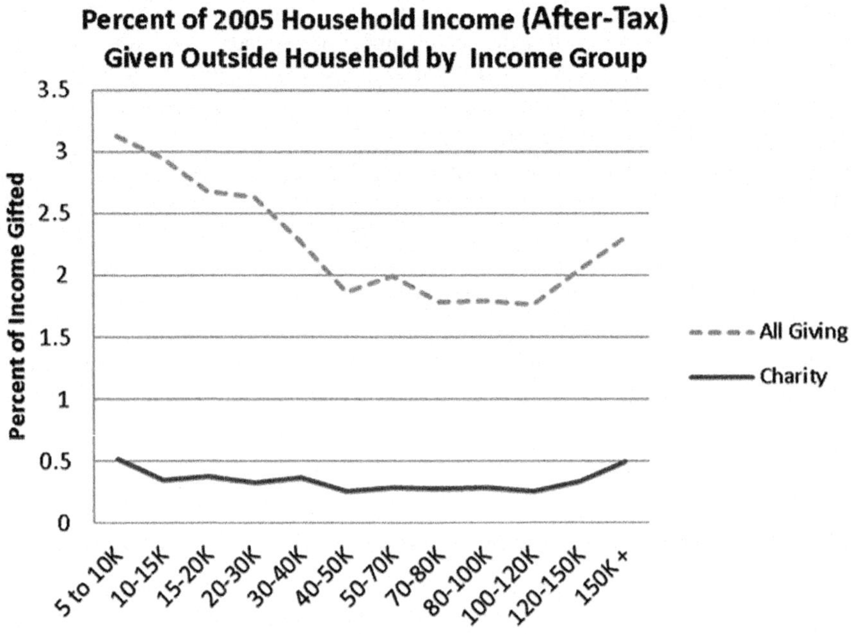

Fig. 5.5 Percent of household income in the United States in 2005 that was given outside the home to all causes (*top, dotted line*) and to charity only (*bottom continuous line*). *Source* Center on philanthropy (See COP 2007)

their lives, they pay more attention to social context, which motivates them to feel more empathic, and ultimately compassionate, for those who share similar experiences of scarcity and suffering (Oveis et al. 2010).

These researchers also found that lower status individuals were more likely than those higher in status to say their values were egalitarian. Furthermore, in playing games, the lower status acted more altruistically and generous than those higher in status. Ironically, higher status persons seem to have high levels of trust generally but feel less concern for the welfare of others than do those lower in status (van Kleef et al. 2008). It appears that lower status persons become more immersed in their social relationships and thus have greater empathy and compassionate caring for the suffering of others, even if it requires a sacrifice (Taylor 2006).

What is perhaps the most intriguing finding from this line of research is that higher income people in some social contexts are more likely than lower income persons to engage in unethical conduct. This tendency has been found in specific situations for cheating, lying, breaking traffic laws, minor stealing, and endorsing minor unethical conduct at work. This tendency is attributed to the upper strata's predisposition toward greed (Piff et al. 2012). Of course, many wealthy and high status people lead exemplary, ethical lives.

Any tendency for affluent people or nations to minimize the problems of those who suffer or live in poverty makes the challenge of alleviating suffering more difficult, In fact, if a society gets to the point where not only are there extreme income differences between the rich and the poor, but the income divide is worsening and the quality of life of the poor is rapidly worsening, it is difficult to imagine resolution of this situation without conflict, violence, and chaos, leaving behind even greater suffering. Most certainly, the suffering of revolutionary change is often far greater than the suffering from moderate sacrifices of resources.

5.4 Inequality and the Widening of Care Divides

The existing global alignment of suffering and compassionate caring (engaged compassion) is the reverse of what is needed for purposes of reducing suffering (and poverty) in the world. When this perspective is combined with data on the flows of charitable aid from developed countries to developing ones, we get a picture of even greater unequal distribution of capabilities to alleviate suffering (GHA 2012).

In 2008, the individuals, organizations and governments of developing countries gave an estimated $169 billion to developing countries (Hudson Institute 2009). This amounts to giving $31 per year to every individual in the developing world. This seems like a sizable amount until you calculate that it is one-fifth of one percent of the Gross World Product. Furthermore, much of the current international aid donations are given explicitly for military purposes, and military efforts often create or extend human suffering, even when they are intended to address suffering.

According to a recent UNICEF (2011) report, the richest fifth of the world population gets 83 % of global income, with just a single percentage point going to those in the poorest fifth. Half of the world's children (1.5 billion under age 25) live in the bottom two-thirds, where the daily wage is $2.00/day or less.

Progress in reducing global inequality is mixed at best. Rises in global inequality have occurred steadily over the past 200 years, but especially in the past four decades. UNICEF, using optimistic assumptions, estimated that it would take more than 800 years for the bottom billion to achieve ten percent of global income. The UNICEF (2011) report argues that "urgency for equitable policies has never been greater." It advocates quick policy actions at national and international levels to ensure a "Recovery for All" that is focused on crushing the forces that push greater and greater income disparities upon us, especially those suffering the most from this social force.

5.5 Inequality, Income Segregation and World Suffering

For many years, scholars and others have warned of the negative consequences of income and wealth-based inequality. As the gap between the rich and poor has continued to spike upward in the past ten years, the warnings continue and the widespread 'Occupy' movement gave the problem much wider visibility.

Epidemiologists Waltzman and Smith (1998), and more recently, Kanachi (2002), highlighted how greater income inequality is associated with higher mortality rates, especially among the very young and older age groups. Their findings are that the effect is generated primarily by the residential segregation caused by income inequality. In the United States, income inequality is intertwined with, and sometimes responsible for, intense racial segregation (Massey and Denton 1993).

Economist Frank (2007) provided evidence that inequality launches 'expenditure cascades' that through social comparison processes perpetuate excessive and compulsive consumption, which drives even greater inequality. Frank argues that not only does this socio-economic process undermine concern for the welfare of others, but it increases vulnerability, financial risk, and unhappiness of those caught in the trap of "keeping up with the Jones."

Wilkinson (2005), documented inequality's negative effect on a number of public health indicators He also assembled evidence that it is associated with lower social trust, lower social capital and civic community, hostility, and violence. Judt (2010) goes further, making a case for the argument that inequality encourages maximization of self-interest and "a kind of de facto authoritarianism." Dowd (2009) addresses related social and cultural impacts of inequality and claims that it goes hand in hand with increased racism and militarism, the latter because social inequality leads to inequality in access to power, opening opportunities for industries like the defense industry to acquire greater power.

The impact of inequality of greatest relevance to suffering and its eradication of suffering is that of increased residential segregation by income. A report by the Pew Research Center (Taylor and Fry 2012) found that the percentage of higher-income households living in wealthier neighborhoods doubled in the last three decades from 9 to 18 %. At the same time, a greater percentage of lower-income Americans live in poorer neighborhoods. Specifically, the concentration of the poor increased from 23 to 29 % in the last 30 years. Among the consequences of this growing isolation of the poor is that it widens the care gap, making it harder for the poor to get social and health services from neighbors, service agencies, or institutions.

The U.S. Department of Housing and Urban Development (HUD) conducted a major field experiment called Moving to Opportunity (MTO). After about 14 years, the results over that period are now being reported (Gennetian et al. 2013). People living in subsidized, public housing in one of six largest cities in the United States were randomly assigned to the condition of receiving a large annual stipend to move to a housing unit of their choice in a low-poverty neighborhood.

This MTO experiment was the largest recent social policy experiment by the United States government. The important findings of this study were that families in the less income-segregated condition were much less likely to feel afraid or unsafe and to experience major depression over the 14 years. In addition, these families were much less likely to be at risk of extreme obesity and diabetes. Finally, the female children in these families significantly improved their learning and test score performance. While the effects within 14 years were not as broad as hoped, the gains were large and significant, and most important, can be attributed to less segregated living.

Suppose it would be possible to scale up this experiment and implement income-integrated living conditions, imagine the possibilities in terms of quality of life and the reduction of suffering. The relocation of 1 million residents of the Dharavi slum in Mumbai or the 4 million living in the Neza-Chalco-Itza slum in Mexico City to mixed income neighborhoods would yield profound changes including a huge reduction in world suffering.

5.6 Implications

The global distribution of compassionate action is out of sync with (opposite to) the distribution of need. Fewer people in high-suffering communities are able to give their time and money to charity, but that does not make them less empathetic and compassionate than those living in places with minimal suffering. International relief and development organizations seek to bring caring into countries with extreme, emergency suffering, but rarely do they leave behind institutions within these countries that address suffering in the long run.

Examining the correlation between indicators of preventable suffering and indicators of compassionate action, such as the amount of charitable giving shows that the relationship between suffering and the amount of charitable giving per person (as a percent of income), is negative, i.e., societies with greater suffering have lower rates of personal giving. This is not a total surprise because in many nations, most households have difficulty reaching minimum levels of subsistence. Nonetheless, the implications of these global patterns of misaligned suffering and compassion have ramifications for humanitarian policies and political actions.

In the next chapter, this will be discussed within ethical frameworks relevant to perceived social responsibility for human suffering. By combining an ethical analysis of social responsibility for all those who suffer worldwide, results of the research have the potential to link generosity with attentiveness to global suffering. Keltner et al. (2010) provide evidence that humans have an instinct to be compassionate and Gilbert (2009) argues that on the basis of recent brain research, the human brain tends to be wired with dispositions to be compassionate Goetz et al. 2010.

The promise to relieve extreme suffering reinforces any potential genetic instincts for empathy and compassion, as well as our socialization in human values and a sense of justice. Thus, there is evidence that humans innately respond negatively to human suffering, no matter how distant. Furthermore, our inner desire to stop suffering is probably more powerful than our learned desire to end poverty, increase life expectancy, or even to improve economic growth.

Even though the physical suffering in the world is many times greater than the physical suffering in the United States, the ratio of US spending for world aid is less than 1 % of total spending compared to 60 % for social services within the United States. Private philanthropy from US donors also goes primarily to US suffering rather than world suffering. The US government and philanthropic organizations tend to give, not in response to the distribution of suffering, but in response to the politics of international relations.

Medical historian Dormandy (2006) concluded that "Pain continues to be the single most useful guide to progress." Although a medical doctor specializing in pain management, his perspective reflects the quality of life perspective. Aggregate pain and suffering are useful measures of a society's progress whether the society has little or highly advanced development. Widespread illness and injury holds a nation back by trapping it in a web of poverty, hunger, and negligible economic productivity.

References

Anderson, R. E. (2012). Caring capital websites. *Information, Communication & Society 15*(5), 479–501.

Armstrong, K. (2011). *Twelve steps to a compassionate life*. New York: Alfred A. Knopf.

Boltanski, L. (1993). *Distant suffering: morality, media and politics*. New York, NY: Cambridge University Press.

CAF (2006). International comparisons of charitable giving. Kent UK: Charities Aid Foundation. Retrieved April 30, 2013 from http://www.cafonline.org/pdf/International%20Comparisons%20of%20Charitable%20Giving.pdf

CAF (2012). World Giving Index 2012: A Global view of giving trends. Kent UK: Charities Aid Foundation. Retrieved April 30, 2013 from https://www.cafonline.org/PDF/WorldGivingIndex2012WEB.pdf

Cassell, E. J. (2004). *The nature of suffering and the goals of medicine*. Oxford, UK: Oxford University Press.

Cohen, S. (2001). *States of denial: Knowing about atrocities and suffering*. Indianapolis, IN: Polity.

Coleman, J. C. (1988). Social capital in the creation of human capital. *American Journal of Sociology*, *94*, 95–120.

COP (2007). *Patterns of household charitable giving*. Bloomington, IN: Center on Philanthropy, Indiana University. (Based on 2005 data from the U. S. Bureau of Labor Statistics Consumer Expenditure Survey).

Dormandy, T. (2006). *The worst of evil: The fight against pain*. New Haven, CT: Yale University Press.

Dowd, D. (2009). *Inequality and the global economic crisis*. NY, NY: Pluto Press.

English, C. (2011). *Civic engagement highest in developed countries*. Gallup World Report. Retrieved May 4, 2013 from http://www.gallup.com/poll/145589/civic-engagement-highest-developed-countries.aspx

Fancher, R. (2003). *Health and suffering in America: The context and content of mental health care*. New Brunswick, NJ: Transaction Publishers.

Farmer, P. (1997). On suffering and structural violence: A view from below. In A. Kleinman, V. Das & M. Lock (Eds.), *Social suffering*. Berkeley: University of California Press.

Frank, R. H. (1999). *Luxury fever: Why money fails to satisfy in an era of excess*. New York, NY: Free Press, Princeton University.

Frank, R. H. (2007). *Falling behind: How rising inequality harms the middle class*. Berkeley, CA: University of California Press.

Gennetian, L. A., Ludwig, J. Maude, T., & Sanbonmatsu, L. (2013). Why concentrated poverty matters. *Pathways*.

GHA (2012) *GHA report 2012*. Somerset, UK: Global Humanitarian Assistance. Retrieved May 1, 2013 from http://www.globalhumanitarianassistance.org/

Gilbert, P. (2009). *The compassionate mind*. Oakland, CA: New Harbinger Publications.

Glenn, E. N. (2000). Creating a caring society. *Contemporary Society 29*(1), 84–94.

Glenn, E. N. (2010). *Forced to care: Coercion and caregiving in America*. Cambridge, MA: Harvard University Press.

Goetz, J. L., Keltner, D., & Simon-Thomas, E. (2010). Compassion: An evolutionary analysis and empirical review. *Psychological Bulletin, 136*, 351–374.

Hudson Institute (2009) *The index of global philanthropy and remittances 2009*. Hudson, NY: Center for Global Prosperity. Retrieved April 30, 2013 from http://www.hudson.org/files/documents/Index%20of%20Global%20Philanthropy%20and%20Remittances%202009.pdf

James, R. N, I. I. I., & Sharpe, D. L. (2007). The nature and causes of the U-shaped charitable giving profile. *Nonprofit and Voluntary Sector Quarterly, 36*, 218–238. doi:10.1177/0899764006295993.

Johansson, S., Leonard, R., & Noonan, K. (2012). Caring and the generation of social capital: two models for a positive relationship. *International Journal of Social Welfare, 12*(1), 44–52.

Judt, T. (2010). *Ill fares the land*. New York, NY: The Penguin Press.

Kanachi, I. (2002). Income inequality and economic residential segregation. *Journal of Epidemiological Community Health, 56*(3), 165–166.

Keltner, D. (2009). *Born to be good: The science of a meaningful life*. NY: W. W. Norton.

Keltner, D., Marsh, J., & Smith, J. A. (Eds.). (2010). *The compassionate instinct*. NY: W. W. Norton.

Kleinman, A. (2011). *Anthropology and cross-cultural mental health: the major questions for future research in global mental health* (Slide presentation). Retrieved January 20, 2013 from http://www.segemi.de/files/kleinman_anthropology_cross-cultural_mental_health_6-2011_1_.pdf

Kraus, M. W., Cote, S., & Keltner, D. (2010). Social class, contextualism, and empathic accuracy. *Psychological Science, 21*, 1716–1723.

Kraus, M. W., Piff, P. K., & Keltner, D. (2011). Social class as culture: The convergence of resources and rank in the social realm. *Current Directions in Psychological Science, 100*, 246–250.

Massey, D. S., & Denton, N. A. (1993). *American apartheid: Segregation and the making of the underclass*. Cambridge, MA: Harvard University Press.

McGaghie, W. C., Mytko, J. J., Brown, W. N., & Cameron, J. R. (2002). Altruism and compassion in the health professions: a search for clarity and precision. *Medical Teacher 24*(4), 374–378.

Nussbaum, M. (1996). Compassion: The basic social emotion. *Social Philosophy & Policy. 13*(1), 27–58.

Nussbaum, M. (2001). *Upheavals of thought: The intelligence of emotions*. Cambridge: Cambridge University Press.

OECD (2011). *Society at a Glance 2011: OECD Social Indicators*, Paris, FR: OECD Publishing. doi: doi.org/10.1787/soc_glance-2011-en.

Oliner, S. P. (2008). *Altruism intergroup apology, forgiveness, and reconciliation*. St Paul, MN: Paragon House.

Oveis, C., Horberg, E. J., & Keltner, D. (2010). Compassion, pride, and social intuitions of self—other similarity. *Journal of Personality and Social Psychology, 98*, 618–630. doi:10.1037/a0017628.

Piff, P. K., et al. (2010). Having less, giving more: The influence of social class on prosocial behavior. *Journal of Personality and Social Psychology, 99*(5), 771–784.

Piff, P. K., Stancato, D. M., Cote, S., Mendoza-Denton, R., & Keltner, D. (2012). Higher social class predicts increased unethical behavior. *Proceedings of the National Academy of Sciences, 109* (11), 4086–4091.

Salvati, A. (2008). *Altruism and Social Capital*. Boca Raton: Universal Press.

Sector, Independent. (2002). *Giving and volunteering in the United States*. Washington, DC: Author.

Sorokin, P. A. (1950). *Altruistic love*. Boston, MA: The Beacon Press.

Stellar, J. E., Manzo, V. M., Kraus, M. W., & Keltner, D. (2012). Class and compassion: Socioeconomic factors predict responses to suffering. *Emotion, 12*(3), 449–459.

Sznaider, N. (2001). *The compassionate temperament: Care and cruelty in modern society*. Lanham, MD: Rowman & Littlefield Publishers.

Taylor, S. E. (2006). Tend and befriend. *Current Directions in Psychological Science, 15*(6), 273–277.

Taylor, P., & Fry, R. (2012). *The rise of residential segregation by income*. Washington, DC: Pew Social & Demographic Trends, Pew Research Center.

UNICEF. (2011). *Global inequality: Beyond the bottom billion—A rapid review of income distribution in 141 countries*. New York, NY: United Nations Children's Fund.

Van Kleef, G. A., Oveis, C., van der Löwe, I., LuoKogan, A., Goetz, J., & Keltner, D. (2008). Power, distress, and compassion: Turning a blind eye to the suffering of others. *Psychological Science, 19*, 1315–1322. doi:10.1111/j.1467-9280.2008.02241.x.

Waltzman, N. R., & Smith, K. R. (1998). Separate but lethal: The effects of economic segregation on mortality in metropolitan America. *Milbank Memorial Fund Quarterly, 76*, 341–373.

Watson, T. (2009). *Inequality and the measurement of residential segregation by income in American neighborhoods*. NBER Working Paper 14908. Cambridge, MA: National Bureau of Economic Research. Retrieved May 8, 2013 from http://www.nber.org/papers/w14908

Wilkinson, R. (2005). *The impact of inequality: How to make sick societies healthier*. NY, NY: The New Press.

Wronka, J. (2008). *Human rights and social justice*. Los Angeles, CA: Sage Publications.

Wuthnow, R. (1991). *Acts of compassion: Caring for others and helping ourselves*. Princeton, NJ: Princeton University Press.

Chapter 6
Ending Preventable Suffering: Ethics and Social Change

Affluent people in Western countries rarely experience or encounter severe personal suffering, even if they consume a steady diet of televised world news depicting violence, death, grief, and pain. This mediated experience has been dubbed 'distant suffering' (Boltanski 1993; Cohen 2001) by those who try to understand the difference between societies with first-hand experience of suffering and those with only technology-delivered snippets. 'Distant suffering' is misleading, however, because the typical television viewer becomes quickly desensitized to the emotional aspects of others' suffering. In societies with heavy consumption of news media, especially television and web-media, it seems easy to become desensitized to horrific images and terrible stories with few flutters of true empathy or compassion.

Professional caregivers suffer from a similar problem, which sometimes is called compassion-fatigue. Paul Farmer (2013) captured the essence of the problem by calling it "anesthesia for the soul." Addressing medical students, he pointed out how easy it is to become too tired to care. Another source of anesthesia for the soul, he notes, is succumbing to the "commodification" of healthcare.

Returning to an ethical framework and practicing empathy and compassion for the suffering of others can help overcome distance, disaffection or desensitization. Unfortunately, it is all too easy to forget that everyone is responsible for the alleviation and prevention of suffering and that it is a primary human obligation, as outlined in the next section.

6.1 Ethical Foundations for the Relief of Suffering

For at least 2,500 years, philosophers have debated the implications of human suffering within the context of ethics and morality. Originally, ethical theories focused on the hedonistic assumption that all people should strive for net pleasure. That is, humans would naturally perform a calculus of pleasure minus pain and suffering, hoping to be left with more of the former than the latter. Mill (1859) extended this

R. E. Anderson, *Human Suffering and Quality of Life*, SpringerBriefs in Well-Being and Quality of Life Research, DOI: 10.1007/978-94-007-7669-2_6,

idea into what he called hedonistic utilitarianism, arguing that the ethical course was to maximize happiness while also reducing suffering. Popper (1956) then introduced negative utilitarianism, with the premise that reducing suffering had a far greater value than boosting happiness. By reducing suffering, the equation would surely lean even further toward net pleasure.

Meanwhile consequentialism, an ethical theory, proclaimed that morality could only be determined by the goodness or badness of the outcomes of behavior. The main opposing moral theory to consequentialism became deontology, which purported that motives were more crucial than consequences in determining right and wrong. The leader of this doctrine, Kant, asserted that, like virtues, intentions were central to good or evil behavior. (This is reflected, of course, when courts consider accidents, crimes of passion, and premeditated crimes differently.) Kant also introduced the concept of duty (or moral obligation) to relieve suffering, alongside other intrinsically good or valuable desires (1780).

Only within the last 30 years has serious attention been given to the ethics of care, beginning with Gilligan (1982). Consistent with the emergence of feminist ethics, she argued that the value of care had been greatly downgraded by gendered thinking and roles. Arguing that Gilligan did not go far enough, Tronto (1993) developed a more precise model of the care process and its role in society. Care ethics, as a value system, deserves further advancement, and the health care fields have struggled to apply it within profit-driven models (cf. Loewy 1991; Edwards 1991). Nilsson's (2012) work on the philosophy of compassion is a very positive development, because it signals serious attention toward compassionate caring within the field of philosophy. Such work will ultimately advance our understanding of the implications of suffering and the use of care to relieve it.

Even with this recent work, arguably the most complete and thorough analysis of the moral philosophy of suffering, *Suffering and Moral Responsibility*, was written in 1999 by Mayerfeld. While he is heavily influenced by deontologists like Nagel (1979, 1986) and Kant (1959), Mayerfeld claims to take a hedonistic perspective on suffering. The difficulty of assigning this influential scholar to one school of thought or another, points out how, in order to adequately frame and make a wide range of decisions regarding suffering, it may be necessary to utilize many or all of the philosophical approaches. Although we may prefer one ethical theory over another, we need to be flexible in applying ethical principles to decisions on how best to reduce suffering in specific contexts.

To return to Mayerfeld, the philosopher initially takes an intuitive position that suffering is obviously bad. Then, by exploring the many dimensions and implications of suffering, he builds a strong case for the primacy of suffering reduction. Mayerfeld asserts that suffering is *so* intrinsically bad that it "gives rise to a prima facie (genuine) duty to prevent it." This duty arises, in his view, from three assertions, one of which is that suffering is intrinsically evil for the sufferer.

For Mayerfeld, suffering does not require a religion, ideology, or philosophy to justify its harmfulness; like murder, its undesirability is a given. To those that argue that suffering is instructive, he replies that suffering does not produce good outcomes on its own. Other forces are required should there be positive outcomes from suffering, but suffering produces its own additional negatives like bitterness and illness.

It is necessary, then, to separate the intrinsic evilness of suffering from its educative and redemptive values. Some positive outcomes Mayerfeld outlines include how, upon suffering, one might become more virtuous or wiser (hence the old tradition of the teacher slapping the student who gets an answer wrong). Redemptive suffering occurs when one 'comes to their moral senses' after suffering, and some assume that anyone committing a grave moral transgression *has* to suffer in some equivalent way if they are to be morally regenerated. But Mayerfeld says that suffering remains an evil in itself; any positive value cannot reside within the suffering itself. Even if the virtue gained is worth the cost, the cost cannot be overlooked. One should consider other, more humane ways to obtain the benefit.

Philosopher Thomas Nagel (1986) also believed it to be self-evident that suffering is bad from an *impersonal* standpoint. So, from both personal and impersonal perspectives, there exists a general duty to relieve suffering.

Given that we have a duty to relieve severe suffering, what accounts for the widespread failure to do so? Mayerfeld offers several possible answers. One is that severe suffering is so ubiquitous, if we were to retain awareness and assume responsibility, it would become debilitating. Consequently, we use fear and other defense mechanisms like distraction and denial (Cohen 2001).

Analysts of 'distant suffering' tend to attribute the callous responses to media depictions of suffering to the medium and the technology itself (Boltanski 1993; Cohen 2001; Ong 2012). Kagan (1989) offers a more nuanced interpretation. She calls peoples' typical response to suffering strangers "paleness of belief," a process that allows the information to register only dimly, without absorption. Thus, when we observe strangers suffering, we see them as stick figures lacking the identities of fully human individuals. Paleness of perception makes it easier to discount or disregard the lives of poor or suffering people in distant places and different cultures. To us, they are the 'other,' not our fully realized counterparts.

6.2 Individual Actions to Relieve Suffering

Chapter 4 estimate of global physical suffering or pain in 2010 put sufferers of physical pain at one billion people—a seventh of the worldwide population is thought to suffer from serious physical pain at any one time. In the USA alone, the IHIS survey in 2010 gave an estimate of 25 million adults (13 % of the national population) enduring either *extreme* suffering. Obviously, some people cannot solve this non-trivial human problem on their own, but much suffering is still preventable and relievable, whether with self-care, professional health care, or the help of informal care-givers.

In an attempt to capture the essence of the underlying dimensions of informal, humanitarian, and compassionate-caring actions, the taxonomy shown in Fig. 6.1 uses three critical components: the amount of the sacrifice (large or minor); social

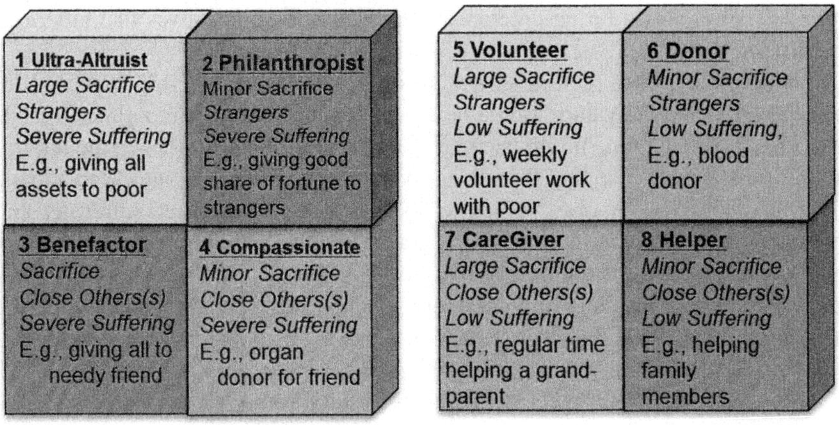

Fig. 6.1 Taxonomy of personal caring and humanitarian actions based upon three defining aspects: sacrifice (large or minor), social distance (strangers vs. close others), and suffering of the other(s) (severe vs. low)

distance (strangers vs. close others), and suffering of the other(s) (severe vs. low). Together these three facets of humanitarianism reveal a continuum from very high to very low humanitarian caregiving. The ends of the continuum are labeled 'Ultra-Altruist' at the high end and 'Helper' at the lowest level.

With its eight types of generosity, the taxonomy suggests that everyone, except perhaps the most extreme sociopath, is a humanitarian, doing things that benefit others, and can be placed along the continuum. Of the three dimensions, only sacrifice is a cost (perhaps in the form of resources, security, or relationships). In fact, Mayerfeld (1999) listed 11 different types of personal sacrifices potentially relevant to actions to reduce suffering (p. 217–218). The other defining attributes within the taxonomy are characteristics of others, namely 'closeness of the relationship' and the 'degree of suffering' of the other(s).

A number of important facets of altruism and compassion are left out of the scope of this taxonomy. One is strong reciprocity, or what is sometimes called 'reciprocal altruism.' By stating that the scope of actions represented encompasses only those actions intended primarily to benefit others, we exclude those actions that are primarily driven by expected reciprocation or that benefit only the helper. The framework also leaves out a distinct representation of empathy, except insofar as the emotion of empathy produces a commitment to the welfare of another.

It is notable that this framework does incorporate a new type of other-oriented action called 'otherish' by Grant (2013). He defines otherish as "being willing to give more than you receive, but still keeping your own interests in sight." The interesting aspect of this concept is that one should give more than one receives. In most affluent societies, this is both a commonly espoused notion but an inconceivable course of action.

The diagram illustrates the continuum and taxonomy by limiting the categorization to only two states of each of the three dimensions. Each of these eight categories is represented by a ranked cell, and these cells are labeled with types of action represented by the associated values of the three variables. For example, box 8, the most other-benefiting action, is labeled 'Ultra-Altruistic' because it represents strangers, severe suffering, and heavy sacrifice. At the other end of the extreme of other-benefiting actions is box 1, which is labeled 'Helper.' These actions are marked by remote others, low suffering, and minor sacrifice.

The remaining cells show the range between these two extremes of humanitarian action. All cells are ranked from 8 to 1, where the highest number cell represents the highest level of compassionate caring, likely to produce the highest level of moral admiration. At the other end of the continuum lies the least amount of compassionate caring, characterized by minor sacrifice, close others, and low suffering.

This illustration is intended to be suggestive of the key forces underlying compassionate caring. It also illustrates the wide variety of actions that individuals can take toward alleviating the suffering of others. It is, essentially, good news. As noted in the beginning of this section, the amount of world's suffering is enormous. By many people tackling it together from many different angles, major dents can be made in global as well as local suffering.

6.3 Institutional Change to Relieve Suffering

Social institutions such as organizations, standards, official policies, and normative practices are needed to implement a radical change in culture. Such reformation is needed because of the embeddedness of destructive elements such as racism and retribution in communities and societies. These destructive cultural elements cause and maintain social suffering. Wilkinson (2013) shows how suffering is both an outgrowth of rationalization in societies and a force that drives social change.

As noted early, social suffering differs from other types of suffering in that it typically is imposed upon one identifiable class of people by another. One classic example of social suffering is genocide, but any stereotyped group (such as the disabled) that experiences social discrimination also feels the negative effects of social suffering. The alleviation of social suffering generally requires support and relief at the institutional level (from government agencies and other human service organizations), because its roots are woven into the fabric of communities and societies.

Relief of social suffering may require major change in social environments. Most of the institutional change needed falls under the banner of social justice, which includes human rights and development. Not only does the relief of suffering require a radical expansion of global health programs, but education in non-violence with training in conflict resolution and reconciliation as well.

Human rights are an anathema for political and social institutions that use autocratic control mechanisms, especially torture. The 'Universal Declaration of Human Rights,' passed by the UN in 1948, remains the principal guidebook for initiatives of human rights and social justice. This document not only states that

"No one shall be subjected to torture or to cruel, inhuman or degrading treatment or punishment," but also outlaws slavery and discrimination. No other social movement tackles social suffering as directly as human rights organizations, which work hand-in-hand with other social justice movements.

According to Schulz (2002), "promoting human rights not only benefits potential victims of the violations, but it also serves the national interest because it showcases a country at its best to the rest of the world." From the standpoint of action to stop suffering, the human rights sector is a critical partner.

6.4 Change in Social Policy for Aid and Welfare Programs

It would appear that many, if not most, public health and other humanitarian initiatives are driven largely by the mission to promote economic growth, to reduce poverty, to improve health, or to prolong and save lives. Reduction of suffering is not an explicit aim. While life, poverty, and health obviously are desirable ends, they do not benefit from the instinct for compassion and the empathic human intuition to alleviate any severe suffering. Disaster relief agencies already capitalize on the human need to relieve the suffering of others. If development aid and welfare programs were to follow suit by appealing to the human desire to avoid suffering rather than touting poverty reduction as their primary goal, the social programs would likely meet with greater public support. But explicitly justifying aid and welfare programs by appealing to the goal of lessening suffering may not be sufficient—it may also be necessary to lay the groundwork for these programs with moral education directed at undermining racism, hatred, retribution, violence, and other culprits of social suffering.

Even among those aid and welfare organizations that identify suffering relief as an implicit objective, little emphasis is typically given to this goal and measuring progress toward it. In the promotional materials of aid organizations like UNICEF, suffering is rarely mentioned, except in the context of reducing starvation from hunger or malnutrition. Public health programs, too, disregard the relief of pain and suffering as a stated goal in the United States, except in the context of hospice and palliative care.

The first step toward giving greater value to lessening suffering would be to assess the typical levels of suffering experienced by those faced with each of a number of different types of calamities. Armed with such information, it would be possible to begin to estimate the cost and feasibility of preventing the calamity or otherwise reducing the prevalence of the suffering. Measurement is a surmountable challenge.

Perhaps suffering has also been neglected in Western societies because many religions cast suffering as a relational good, a sign of character strength. In recent years, the field of pain medicine has been accused of fostering drug addiction by overprescribing opiates. The message is that potential addiction is worse than certain pain. This moral judgment and risk assessment may have resulted in the reduced use of potentially addictive medications alongside a negative attitude toward both the prescriber and the user of potentially addictive drugs for pain management. The greatest tragedy is that now doctors in the United States have come to withhold pain-relieving medications even when patients are at low risk of addiction and are experiencing legitimate suffering. More research needs to be done on these issues to determine public knowledge and behavior from the prescription of pain medications to opinions about the relationship between suffering and character-building.

6.5 Implications for Quality of Life Research

Definitions and indicators of the quality of life have generally focused on the antitheses of suffering. An intuitively appropriate approach, this metric is undermined by the findings summarized in this book. Here, by focusing on suffering and related negative dimensions of social well-being, we have gained insight into the quality of life from previously neglected perspectives.

Neither the research community working on quality of life nor those working on well-being and happiness has addressed their opposites. Equally important to the human condition of well-being are ill-being and evil-being; happiness as well as unhappiness; quality of life as well as quantity of squalor; and life satisfaction as well as life dissatisfaction.

Quality of life, without a doubt, is multidimensional, involving both positive and negative elements. If respondents are asked only about the positive elements, the overall portrayal may well be biased or incomplete. The same is true for asking questions only about negative elements.

Part of the challenge is that some people compartmentalize suffering, not allowing it to shape their assessments and actions. Also, there may be cultural pressures to be optimistic and upbeat rather than totally accurate. Suffering and pain need to be measured along with positive elements until we understand more about the effectiveness and validity of different types of questions to elicit states of well-being and quality of life.

In the same way that both compassion and self-compassion are needed for balanced adjustment to external demands, so do individuals need to be concerned for other-oriented QOL (quality of life) and self-QOL. One can think of QOL as an attribute of both self and others. Separately measuring these two types of QOL offers a useful extension to traditional QOL research.

6.6 Cutting Edge, Contentious Issues Related to Suffering

The issues presented here include points of view that are in flux worldwide. They are likely to receive future attention, and considering them now will help us anticipate upcoming contention and refocus on the problem of suffering.

6.6.1 The Relief of Suffering Versus Obligation to Sustain Life

Medical advances prolong life and reduce suffering. They can also prolong suffering. This creates a dilemma: some people must now choose between keeping a suffering person alive or 'putting them out of their misery.' This dilemma has already formed the foundation for a powerful right-to-die movement and a counterbalancing right-to-life movement. The contentious debates and court battles between these two sides have become politicized. Health care services to keep those who are in agony or have lost consciousness come at high cost to society (Ekland-Olson and Aseltine 2012), But terms like 'mercy killings' and 'assisted suicide' have not helped in the development of policies and laws that address either the desires of individuals or the needs of society.

Euthanasia has been discussed for 150 years in the United States, and, since the 1980s, there has been one major court case after another. What has not emerged is a clarification from moral philosophy of the importance of the duty to relieve suffering. As a growing number of people take the duty to reduce suffering seriously, it might be possible to work out compromises that both honor the goal to minimize suffering and allow individuals and their families to guide the process.

6.6.2 The Relief of Suffering Versus Protection from Addiction

A somewhat similar issue exists around the importance of relieving sufferers versus protecting them from addiction. It is commonly believed that morphine and other opiate pain relievers can quickly become habit forming, so there is growing pressure in some countries to greatly reduce and impose controls on prescriptions for these drugs. Melzack (1990) and Taylor (2007) provide evidence that much public opinion and many policy-makers' views on the risks of addiction are ill-informed. Large clinical studies have found that without previous substance abuse, there is a near zero chance of contracting an addiction from the doctor-supervised use of narcotics for pain-relief. Furthermore, those with histories of drug abuse usually have symptoms of depression and other more severe psychological disorders, making it is possible to screen for such conditions and treat pain without serious risk (Melzack 1990).

As noted by Melzack, the irrational restrictions against widespread use of narcotics in the treatment of pain most harm children and the elderly. Too often, children with temporary but very severe pain do not receive adequate relief. And the elderly suffer for much longer periods than younger adults when recovering from surgery or injury. Consequently, they suffer needlessly and more when they are denied pain-killers. Those who lobby for imposing greater controls tend to be people that place little value on the reduction of suffering. Those who fight against such controls do not frame their arguments with sufficient emphasis on the human duty to relieve suffering. The argument against suffering thus tends to lose within courts—and the court of public opinion.

6.6.3 Suffering Relief Versus Economic Relief

As noted above, advocates of humanitarian assistance to global sufferers tend to organize their programs and policies around economic aid and other assistance not directly related to suffering. Because of the poor results of these programs, many organizations have been shifting their goals and policies to social and environmental causes such as poverty, water supply, and adequate sanitation (Dowd 2009; Wilkinson 2005a, b).

One basic problem with economic relief alone has been that it exacerbates inequality, actually worsening problems of poverty, health, and needless suffering. Wilkinson (2005a, b) has assembled evidence that rising inequality in most countries has worsened health, lowered social trust and social capital, and may have even increased hostility and violence.

Based on this evidence, major development organizations such as UNICEF and the World Bank have, to some extent, shifted their priorities to poverty and health (UNICEF 2011). This is not enough: inequality continues to grow, threatening loss of trust, community, and civic organization (Dowd 2009).

If global and aid-donating nations were to shift their policies to offering suffering relief programs as their major approach, combining their resources with programs that coordinate personal helpers like Peace Corps members, the resulting global solidarity and willingness to collaborate might do wonders for development and well-being.

6.6.4 Accountability and Responsibility for Suffering

The human rights movement has made great strides in establishing the principle of harm-doers being accountable for the suffering of their victims. Harm-doers that cause suffering may be individuals, groups, or governments of any kind. While many governments have systems of justice that administer criminal punishment to individuals as a means to hold them accountable for the suffering of their victims, little progress has been made toward holding organizations and governments responsible for suffering created by their policies and practices. Often when an organization is taken to court, the sentence received is not commensurate with the human suffering produced by their actions. This may be a sign that appropriate laws are not in place or that politics has produced a biased, unfair justice system.

Of even greater concern is that neither nations nor their governments are held accountable for the degree of suffering produced. This is especially problematic in times of armed conflict or war. For example, in response to the 9/11 atrocity in which about 3,000 people were killed, the United States declared a war on terror and invaded several middle-eastern countries. The total record of deaths and other suffering is still accumulating, but many thousands of people have been killed, millions have been displaced from their homes, and many additional people still suffer from injuries, illnesses and grief.

At least two major weaknesses in the world system make such huge magnitudes of suffering possible. One is that the international justice system is not able to impose sanctions on nations, especially powerful ones, for producing needless suffering, and the second is that no one requires accountability and good record keeping on casualties of armed conflict. Since it is advantageous to the suffering-producers to retain secrecy on the number of casualties of various types, it becomes impossible to fully assess the amount of suffering left behind.

The suffering produced in armed conflicts is often justified by the need for safety of one's citizens. Ironically, all too often the perpetuation of conflict yields less safety, rather than more. Resolving these conflicts, like the reduction of violence, is not straight forward, nor is it easy. What could help reduce suffering in future conflicts is insisting that people and nations have a moral and legal responsibility to prevent as much future suffering as possible.

6.6.5 Is Human Progress Possible without Major Strides in Relief of Suffering?

In the past 10–15 years, major progress has begun on global and national indicators of human progress. A major element in this progress has been widespread recognition that economic growth and GDP are not sufficient measures of human progress. The UNDP (United Nations Development Program) publishes an annual Human Development Index that incorporates health and education into what is otherwise an income-based index. The OCED (Organization for Economic Co-operation and Development) started the 'Better Life Initiative,' which offers a participatory tool: anyone can insert their own priorities for human progress in constructing a model index that uses about 10 different components in addition to income.

In April 2013, a brand new index was released by the Skoll Foundation and four other private foundations (Social Progress Index 2013). The new Social Progress Index is a composite of 52 indicators across 3 major dimensions of progress: basic human needs, well-being, and opportunity for people to reach their full potential. While, on the surface, this index would seem to be a major step toward addressing the needs of all human beings, from the basic to the creative, it does not go far enough. Only a few indicators reflect the degree of social suffering (e.g., undernourishment and deaths from cancer, HIV, heart failure, and diabetes). So much more could be done in collecting both subjective and objective indicators related to the intensity of suffering.

On the one hand, progress is being made in both global and local policy making and indicator development around human progress for all. On the other hand, the progress appears to be minor at best. Hopefully, the policies and their positive impacts will grow exponentially, but the odds are against it. Inequality and self-interest appear to be rising, and there is little basis for optimism. That could change rapidly, of course, particularly if new generations refuse to numb themselves to the horror of the suffering experienced by their fellow human beings. Opening themselves up to empathy, compassionate caring, and sacrifice, our protégés could keep the human family from self-destructing.

Watch the measures of human progress that are developed and refined in the coming years. What happens will reflect how quickly global institutions move toward recognizing the constraints of progress that can only be loosened with the recognition of the great value of every human being and the agony of those struggling under severe suffering. A successful future depends, in part, upon how successful global society is in recognizing and alleviating human suffering everywhere.

6.7 Conclusion

Suffering unfolds an array of deeply human ironies. Every major religion calls for compassion and aid for others who suffer, yet the number burdened with severe suffering continues to expand. Those who reach out to others suffering may themselves suffer, but many feel joy from having reduced someone's suffering.

The sheer volume of global suffering will continue to rise for many years into the future due to higher rates of population growth in the *global south* and increasing longevity (UNDP 2013). Likewise, environment changes suggest that future disasters will sharply increase in severity and suffering. Meanwhile, attention to suffering as a topic or issue has been declining. Von Wiese in 1934 said that suffering is the "fundamental problem of sociology," yet today it is rare to find an article, much less, a book on the subject by sociologists. The U.S. Congress Committee on Foreign Affairs in 1921 held a hearing on "Relief of suffering populations of the world," yet in the past 50 years, the word 'suffering' has rarely been used in political rhetoric in contemporary American politics.

Another irony is that the powerful contemporary institutions established to ostensibly reduce suffering primarily address poverty and economic development rather than suffering. While economic resources can help reduce suffering, they may also increase suffering by increasing inequality *and* expectations. Perhaps the biggest tragedy is that in an age of globalized media, those who hold charitable resources have become largely desensitized to horror and suffering, especially when it lies outside their neighborhood or national boundaries (Boltanski 1993; Cohen 2001). Suffering statistics, as compared to poverty statistics, have more potential for arousing public interest and mobilizing action to improve the conditions of those in severe suffering, but we must also encourage empathy and identification.

Fundamentally, human suffering is the greatest humanitarian challenge today. Suffering may be a consequence of hunger, poverty, violence, illness, injury, or depression, and it directly incapacitates people through fear and physical immobility. Suffering also generates social disorder, threatening the survival of individuals, communities, and societies. Genocide epidemics are the most vivid examples of the horror of mass suffering and the tragic chaos of decaying societies. Such mass suffering has become a measure of social disorder and decay, analogous to a high body temperature that signals a serious threat to the life of the individual. The claim has been made time and time again throughout this book that a successful future depends, in part, on how successful global society is in recognizing and alleviating human suffering everywhere. Suffering is not just a matter of humanitarian concern: the reduction of suffering should be included in effective strategic planning for global social and economic progress.

Taking on the suffering of others is, essentially, the development and operationalization of compassion. But we cannot become truly compassionate toward others without also being compassionate to ourselves. Alleviating suffering is a force for healing the world—and ourselves.

References

Boltanski, L. (1993). *Distant suffering – morality, media and politics*. New York: Cambridge University Press.

Cohen, S. (2001). *States of denial: Knowing about atrocities and suffering*. Indianapolis: Polity.

Dowd, D. (2009). *Inequality and the Global economic crisis*. New York: Pluto Press.

Edwards, B. S. (1991). An ethics of suffering: Does it solve the problems we want it to solve. *The Journal of Clinical Ethics, 2*(2), 107–112.

Ekland-Olson, S., & Aseltine, E. (2012). *How ethical systems change: Tolerable suffering and assisted dying*. New York: Routledge.

Farmer, P. (2013). *To repair the world*. Berkeley: University of California Press.

Gilligan, C. (1982). *In a different voice*. Cambridge: Harvard University Press.

Grant, A. (2013). *Give and take – A revolutionary approach to success*. NY: Penguin Group.

Kagan, S. (1989). *The limits of morality*. Oxford: Clarendon Press.

Kant, I. (1959). *Foundations of the metaphysics of morals*. Translated by L. W. Beck. Indianapolis: Bobbs Merrill.

Loewy, E. H. (1991). *Suffering and the beneficent community: Beyond libertarianism*. New York: State University of New York Press.

Melzack, R. (1990). The tragedy of needless pain. *Scientific American, 26*(2), 27–33.

Mill, J. S. (1859) *On liberty*. Reprinted in 1975 by Norton, New York.

Nagel, T. (1979). *Mortal questions*. Cambridge: Cambridge University Press.

Nagel, T. (1986). *The view from nowhere*. New York: Oxford Univ. Press.

Nussbaum M.C. (2011). *Creating capabilities—The human development approach*. Cambridge: Belknap Press of Harvard University Press.

Ong, J. C. (2012). 'Witnessing' or 'mediating' distant suffering? Ethical questions across moments of text, production, and reception. *Television New Media, 24*(1), 1–18.

Popper, K. (1956). *The open society and its enemies*. Princeton: Princeton University Press.

Schulz, W. (2002). *In our own best interest: How defending human rights benefits us all*. Boston: Beacon Press.

Social Progress Index. (2013). Social Progress Index 2013. Washington DC: Social Progress Initiative. Retrieved May 15, 2013, from http://www.socialprogressimperative.org/publications

Taylor, A. L. (2007). Addressing the global tragedy of needless pain: Rethinking the United Nations single convention on narcotic drugs. *Journal of Law, Medicine and Ethics, 35*(4), 556–570.

Tronto, J. C. (1993). *Moral boundaries: A political argument for an ethic of care*. New York: Routledge.

U.S. Congress Committee on Foreign Affairs (1921). *Proceedings of a Hearing On "Relief Of Suffering Populations Of The World, Stricken By War, Famine, And Pestilence*." Washington DC: Government Printing Office. Reprinted in 2012 by Forgotten Books. Retrieved May 22, 2013, from www.forgottenbooks.org.

UNDP. (2013). *Human development report 2013 - The rise of the south: Human progress in a diverse world*. New York: United Nations Development Programme.

UNICEF. (2011). *Global inequality: Beyond the bottom billion - A rapid review of income distribution in 141 countries*. New York: United Nations Children's Fund.

von Wiese, L. (1934). Sociology and Suffering. *International Journal of Ethics, 44*(2), 222–235.

Wilkinson, I. (2005a). *Suffering: A sociological introduction*. Indianapolis: Polity.

Wilkinson, I. (2005b). *From the sociology of risk to a critical sociology of suffering*. Presentation at the SCARR Network Conference. Retrieved March 9, 2013, from http://www.kent.ac.uk/scarr/events/finalpapers/wilkinson.pdf

Wilkinson, I. (2013). The problem of suffering as a driving force of rationalization and social change. *The British Journal of Sociology, 64*(1), 123–141.

About the Author

Ronald (Ron) Anderson is Professor Emeritus at the University of Minnesota. He received his Ph. D in sociology from Stanford University in 1970. From 1968 until retiring in 2005, he served on the faculty of sociology at the University of Minnesota. Throughout that time, he consulted for many government agencies and corporations on survey research and technology-related issues. From 1990 to 2005, he coordinated several international studies of the social and learning effects of information technology within primary and secondary education in 20 or more countries in each study. From that and earlier work, he wrote or edited seven books and over 100 articles. Since retirement, his research interests have focused primarily on compassion and suffering. Further details on his work can be found in the following websites:

en.wikipedia.org/wiki/Ronald_Anderson
www.soc.umn.edu/~rea/
StopSuffering.net.

R. E. Anderson, *Human Suffering and Quality of Life*, SpringerBriefs in Well-Being and Quality of Life Research, DOI: 10.1007/978-94-007-7669-2, © The Author(s) 2014

Index

R. E. Anderson, *Human Suffering and Quality of Life*, SpringerBriefs in Well-Being and Quality of Life Research, DOI: 10.1007/978-94-007-7669-2, © The Author(s) 2014

CPSIA information can be obtained
at www.ICGtesting.com
Printed in the USA
LVOW04s1038190117
521520LV00011B/178/P

9 789400 776685